Contents

Preface

This book treats estimating and tendering as a major management function within construction companies. It describes the process of producing an estimate and a tender, the personnel involved and presents the calculations and decisions involved with examples and typical data. The second edition updates these calculations and data. In the first edition the major change then facing working practices was that of the introduction of computers into estimating. Since then much experience has been gained in computer usage and computer systems have continued to evolve. The second edition reports on the typical systems now available to estimators by way of computer aids. The management of implementing computer systems and the training of staff still remain issues faced by many and these are addressed. A new chapter describing the additional issues faced in overseas estimating has been added.

It is hoped that this second edition will continue to appeal to established senior estimators for its information relating to new technology and as a training aid for their junior staff, and to young construction executives who are considering a career in estimating and who wish to appreciate the process of estimating and tendering and the related calculations.

It is pleasing that demand for this text has increased from students and lecturers on both undergraduate and postgraduate courses in civil engineering and construction management. It is evident that the nature of undergraduate courses in changing rapidly and now includes more commercial subjects such as estimating and tendering.

Acknowledgements

The authors would like to extend their sincere thanks to all those who have provided contributions, advice and assistance from which this book has benefited. In the writing of this new edition the authors are particularly grateful to those who have assisted with the new and revised chapters.

Philip Oliver and Ron Randell provided considerable advice and assistance with regard to the chapter on estimating and tendering for overseas works.

The chapter on computer aided estimating would not have been possible without the assistance of Tony Day, Stephen Thompson and their colleagues at Digital Building Systems Ltd. Even though the writing of the text coincided with the launch of the new 'Conquest' estimating system, they found the time to advise and assist us in ensuring that the latest aspects of computer aided estimating were included.

In the production of the book the authors are grateful to Vera Cole for word processing the text, to Janet Redman for drafting the diagrams, and to Sherif Oteifa who spent numerous hours checking the arithmetic, proof reading the text and updating the reference list.

To all others who provided the contributions, advice and assistance from which the book has benefited we extend our sincere thanks.

BBEST is the registered name of the estimating system developed by Balfour Beatty Ltd. Conquest is the registered trade name of the Conquest Estimating System developed by Digital Building Systems Ltd. UNIX is the generic name given to the operating system originally designed by AT and T Bell Laboratories Inc.

1 Civil Engineering Works and Estimating

INTRODUCTION AND AIMS

The work of construction contractors can be crudely represented as:

(a) getting work at the right price; and

(b) doing the work within that price.

Much of the initial glamour and attraction of a career with a civil construction contractor is in 'doing the work'. The equally important task of obtaining the work, estimating and tendering is at first sight less appealing to many young construction executives. The estimating department has however provided an exciting and rewarding career for many commercially minded engineers.

Much of the excitement is derived from the importance of estimating and tendering to the growth and prosperity of the company. If a company fails to win contracts or wins contracts with an inadequate price its future is in jeopardy. The work of an estimator is more than simple calculation of costs: it is also a management function because the process of producing an estimate and a tender is complex, involving many different tasks and personnel within the company and external organisations such as the design staff, suppliers, sub-contractors and plant hire companies. This process has to be managed within a tight and well defined time frame: the tender deadline must be met. The excitement is heightened by the uncertainty: was our tender the lowest?

The work of estimators is also changing because like many other business functions the intrusion of computers progresses relentlessly. So estimators are having to adjust their work practices and procedures in order to take advantage of the facilities that computers can offer.

The aims of this book are:

* to describe the process of estimating and tendering for both the UK and overseas;
* to explain and give examples of the calculations in estimating and tendering; and
* to introduce the use of computers in estimating and describe the issues faced in implementing such systems.

The intention is to progress from the overall process through the calculations and end with descriptions of the computer aids and the problems of implementation.

CONTENTS

In fulfilling these aims each chapter has a specific function and these are:

Chapter 2.   The estimating and tendering process within a contractor's organisation.

Estimating and tendering is not always viewed as a process and this chapter describes the process from receipt of invitation to tender to the submission of a tender, the personnel within the contractor's company and the external organisations involved in the process.   The stages of the estimating and tendering process and their interrelationship are described in detail.

Chapter 3.   The calculations in the estimating process.

The principles of the calculations used in estimating and tendering such as 'all-in' labour rates, plant rates,   material quotations and costs, and the calculation of direct cost rates by 'operational' and 'unit rate' methods are presented.

Chapter 4.   Planning and estimating.

In civil engineering work planning is an intrinsic part of the estimating process whether it is undertaken by the estimator himself or by a planner acting in support of the estimator.   This chapter describes the planning techniques employed by estimators including bar charts and network analysis.   The relationship between estimators and planners is also described.

Chapter 5. Examples of the calculations within the direct cost estimate. Examples of direct cost estimating using the principles of the calculations explained in Chapter 3 are given for: earthworks; drainage; formwork; reinforcement; concrete and overheads. Typical data used by the estimator are also presented. These data are not intended to be comprehensive.

Chapter 6. The calculations and decisions in tendering.
The distinction between 'estimating' and 'tendering' will be made later in Chapter 1. Whereas Chapter 5 deals with 'estimating' examples, Chapter 6 deals with examples of the decisions and calculations in tendering. Included are descriptions of the information and reports required for the tender meeting and examples such as resource (labour, plant and material) summaries, cost summaries and calculations relating to cash flow, overheads and profit.

Chapter 7. Estimating and tendering for overseas work.
The estimating and tendering process for overseas work is compared with the process for estimating and tendering in the contractor's own country. The specialist areas where the contractor needs to seek expert advice are indicated, together with the implications of the additional risks involved.

Chapter 8. Estimating and bills of quantities.
In the past a major impediment to effective use of computers in estimating has been the form of the bill of quantities. The advent of the Civil Engineering Standard Method of Measurement (CESMM2) and the Standard Method of Measurement for Building Works (SMM7) has done much to alleviate this although problems still remain in the Department of Transport's Method of Measurement for Highway Works and in earlier versions of building bills which are still in use. The link between estimators' data and bill items, prepared under these methods of measurement, is explained together with the notations required to achieve it.

Chapter 9. Computer aided estimating
The use of computers in a contractor's organisation is reviewed together with a detailed examination of the use of computers in estimating. The difficulties in adopting computer aided estimating are discussed together with the range of hardware and types of computer system currently available. The four phases of implementing a computer system within a contractor's office are

described together with the support requirements of the contractor's staff. Future directions in computer aided estimating including the impact of expert or knowledge based systems are discussed.

The remainder of Chapter 1 is intended to establish:

* the contractual framework within which the estimator works with respect to the promoter of the works by briefly considering the method of obtaining tenders and types of contract;
* the broad make-up of an estimate and tender, including the differences between estimating and tendering and the importance of the estimate in a tender;   and
* the evaluation of the tender by the client's advisers.

METHODS OF OBTAINING TENDERS AND TYPES OF CONTRACT

Estimators are constrained to work within the overall tender and contractual procedures of the industry.   The Institution of Civil Engineers' Civil Engineering Procedure (9) reviews the methods of obtaining tenders and the types of contract used in civil engineering work.   For detailed descriptions reference should be made to this document.

Tenders

The methods of obtaining tenders are, in broad categories:

* selective tendering;
* open tendering;   and
* negotiation.

Data from ten UK contractors revealed that over 80% of tenders submitted were through a selective tendering procedure.

Selective tendering is practised by selecting a 'short list' of contractors to be invited to tender, usually from a previously compiled list of approved contractors or from special lists created for particular contracts.

Contracts

The Civil Engineering Procedure lists the following types of contract:

* measurement contracts

> (i)   bill of quantities;
>
> (ii)   schedule of rates;

* lump sum;

* cost reimbursement;

* all-in contracts.

A survey of ten contractors revealed that over 80% of tenders submitted were for measurement contracts based on a bill of quantities.

Contract documents

The contract documents in a measurement contract based on a bill of quantities are:

* instructions to tender;

* form of tender, for example see pages 33 and 34 of the 'ICE Conditions of Contract, fifth edition';

* Conditions of Contract, for example 'ICE General Conditions of Contract, fifth edition';

* specification, for example see 'The National Water Council's Civil Engineering Specification for The Water Industry' and 'The Department of Transport's Standard Specification for Highway Works';

* bill of quantities, prepared by a standard method of measurement, for example the 'Civil Engineering Standard Method of Measurement CESMM second edition' and 'Department of Transport Method of Measurement for Highway Works' or the 'Standard Method of Measurement for Building Works';

* form of agreement, for example see page 35 of 'ICE Conditions of Contract, fifth edition'; and bond.

These documents define the constraints within which the estimator must work and the information to be supplied with the tender. The document that influences the form of the estimator's calculations is the bill of quantities because it determines the form of most of the data to be presented in the tender submission. Fig. 6.9 in Chapter 6 gives an example of a bill of quantity page prepared for submission.

ESTIMATING AND TENDERING

This section describes the broad make-up of an estimate and tender.
Tenders can be seen as being made up of three elements:

* direct costs;

* on-costs;   and

* mark-ups.

Direct costs

Direct costs are predominantly the labour, plant, materials and sub-contractor costs involved in executing the works.   The direct costs have to include allowances for materials handling and wastage, plant idle time, erection and dismantling of temporary works, such as scaffolding and falsework, temporary works such as piling, and consumable materials not built into the permanent works.   Chapters 3 and 5 explain the calculations involved and give examples.

On-costs

On-costs are costs which can be directly attributable to a contract and broadly include: site staff; transport costs; welfare and site office costs; insurances and bonds.   Chapter 5 describes these in more detail.

Mark-ups

The direct costs and the on-costs together represent the costs to the contractor in undertaking the contract. It is the responsibility of a contractor's estimating staff to establish these costs. The direct costs plus the on-costs are called the 'cost estimate'. The tender total presented in the tender is a contractor's 'selling price'. The difference between the cost estimate and a contractor's selling price is the mark-up.

The mark-up usually contains three elements: an allowance for company overheads, an allowance for risk and an allowance for profit. There is a fourth element that is part of the consideration of the total tender sum and that is the discounts obtained, or expected to be obtained on material purchases and sub-contracts. If these discounts have not been deducted the 'cost estimate' already has an allowance included. When considering what further allowances for overheads, risk and profit to add this fourth allowance will be evaluated.

The evaluation of the mark-up and the determination of how the costs and mark-up will be allocated to the bill items is the part of the process described as tendering and the decisions involved and examples are given in Chapter 6.

THE IMPORTANCE OF THE ESTIMATE

A contractor's tender is based on an estimate of the cost to the contractor of constructing the works. The measure of a successful tender is one whose value is low enough to win the contract and high enough to allow the contractor to discharge the responsibilities defined in the contract and still show some profit. In any tender as much as 90% to 95%, or even more, of the tender sum is represented by the estimated costs, the remainder being the contractor's mark-up to cover profit and company overheads. Whereas much is written on theories of bidding and the analysis or probability of winning against mark-ups the proportion of the tender sum made up by the estimated costs makes it obvious that the major factor in determining a tender's success is the cost estimate.

DISTINGUISHING THE ESTIMATE FROM THE TENDER

Chapter 5 will deal with *direct cost* estimating examples and Chapter 6 will deal with *tendering* giving examples. While in all companies the estimators are party to the tendering decisions and, indeed, in some companies have sole responsibility for tendering, the difference between *estimating* and *tendering* needs to be borne in mind so that the calculations and decisions can be undertaken with the required clarity and accountability. The primary function of the estimating staff is to prepare the estimates of the *direct cost* and *on-costs*.

TENDER EVALUATION

The climax of the estimating and tendering process is the submission of the tender. In some cases where the contract is based on a bill of quantities, this is preceded by entering the item rates in the bill of quantities, calculating each item sum, the page, section and bill totals.   Where the contract documents comprise more than one bill the sum of all bill totals represents the tender total. The tender total is thus made up of the total of all bill item sums. This tender total, therefore, represents to the contractor all the costs, both direct and indirect, together with additions included to cover profit, risk and company overheads, often called mark-ups. The client's advisors' staff will have been responsible for preparing the contract documents, inviting the tenders, attending site visits and answering queries.

In preparing a report on the tenders received the evaluation by the client's advisors will comprise:

* technical evaluation of any alternative designs or proposals;
* consideration of any qualification, if allowed, included by the tenderers;
* checking of arithmetic of the tenders;
* examination of rates quoted for work in order to identify:
  - unrealistic rates;
  - rate loading;
  - tactical pricing aimed at taking advantage from possible design or quantity changes; and
  - errors or omissions.

The rates of, say, the three lowest or sometimes all tenderers are usually tabulated for comparison.   This means that the rates for each contractor are being cross compared. The main source of the client's advisors cost data for these comparisons and for estimating the cost of future projects is from priced bills of quantities. The value of these data for such purposes is suspect and can be undermined by the adjustments made by contractors in entering rates against the bill items. One such adjustment is 'rate loading', a device of increasing rates on certain items of work and decreasing them on others thereby maintaining overall a competitive tender while improving the contract's cash flow or enhancing the returns from price adjustment clauses. Such rate loading may be shown up by comparison but if all tenders indulge in similar tactics to broadly the same degree such devices cannot be identified by comparison. The range of rates against bill items is much wider than the range of tender sums for the whole contract. Thus the cross comparison of individual item rates is to some extent subjective. However, the client's advisors' staff is charged with the responsibility of scrutinising the tenders and recommending which tender should be accepted.

2 The Estimating and Tendering Process Within a Contractor's Organisation

INTRODUCTION

The estimating and tendering process within a contractor's organisation takes place between two well defined points, the decision to tender and the submission of the tender. Between these two there is a complex process that involves and depends on several different categories of personnel within the contractor's organisation as well as essential external contributions by the client's advisors, materials suppliers, plant hire companies and specialist sub-contractors. The process includes collecting and calculating cost data, deciding on and planning around a construction method, selecting resources and production or output rates, combining costs and resource usages to determine cost estimates, calculating site on-costs, taking tender decisions and preparing the tender documents for submission.

The involved nature of the process and the reliance on different personnel and outside organisations for data and information require that the process is carefully managed.

This chapter describes the estimating and tendering process together with the personnel involved.

THE PERSONNEL

The personnel involved in estimating and tendering can be divided into three classes:

    (1)   the client's advisors;

    (2)   the contractor's personnel, including senior management, estimators, buyers, plant managers, temporary works designers,

in-house designers, site management staff; and

(3) the external organisations such as material suppliers, plant hire companies and sub-contractors.

The contributions of each of these are as follows:

(1) The client's advisors' staff

The person or organisation for whom the work is to be constructed is normally referred to as the promoter. The promoter will appoint professional advisors for the project, who may be from within his own staff or from a consulting engineering practice. The function of these advisors includes the development, design and technical direction of the works as well as the preparation of specifications, bills of quantities, drawings and other contract documents. It is these contract documents which describe the works to the contractor. The drawings, bills of quantities and specifications are the main sources of information to the estimators who prepare the cost estimates and tenders. The method of measurement used in the U.K. for preparing the bill of quantities is mostly the Civil Engineering Standard Method of Measurement, second edition or the Method of Measurement for Highway Works, or variations of these for civil engineering works or the Standard Method of Measurement for Building Works, seventh edition or earlier versions for building works.

(2) The construction contractor's personnel

Senior management is an expression used to imply company directors and those who hold responsibilities similar to directors. Senior management is usually involved in the decision whether or not to tender for a particular contract and in the decision on the tender to be submitted after consideration of the estimate of cost and resources involved as produced by the estimators. Most companies have a director responsible for the estimating department or an estimating manager who reports to a director. Thus the day to day activities of the estimating department are closely monitored by senior management.

Estimators are the personnel employed in the estimating department charged with the responsibility of producing the estimates and managing the process described in this chapter. Senior estimators are usually professionally qualified staff with extensive experience of the construction industry. These senior estimators are normally supported by junior estimators who may be aspiring to a career in estimating or may be gaining experience in estimating as part of a wider career development.   Other support staff include estimating clerks and computer operators.

Planners are the personnel employed to produce construction plans or programmes. As far as the estimators' requirements are concerned this means the pre-tender programme, which may not be as detailed as one produced for site use but will provide the overall duration of the project and the duration and sequence of the key activities and approximate resource totals for labour and plant. In some companies the estimators produce the pre-tender programme themselves, whereas in others the planners produce it; in the latter case, close liaison is required between the estimators and planners. In some companies the planners are part of the estimating department and in others there is a separate planning department.

Buyers are usually responsible for purchasing materials and placing orders with plant hire companies and sub-contractors. The service they give estimators is to provide quotations for materials, plant hire and sub-contractors. In some companies the estimators are given the task of sending out the enquiries and receiving the quotations.

Plant managers are responsible for the company's plant department and supply estimators with current internal hire rates and advice on likely availability of company owned plant.

Temporary works designers are responsible for designs of major temporary works such as bridge supports or falsework and in civil engineering contracts the estimators would take their advice on the nature and likely cost of major temporary works.

In-house designers are used by estimators when it is decided to submit an alternative design. For example a contractor may decide to submit a tender for a bridge based on prestressed concrete as described in the contract documents and also submit an alternative tender for a steel bridge which would be designed in-house. The outline design work would take place during the tender period and the detailed design and working drawings would be produced later. For the outline design and quantities, the estimators would rely on in-house designers or a consulting engineering practice engaged specifically for the task.

Site management staff are the personnel who are employed to take responsibility for the execution of projects on site. This expression covers agents, works managers, engineers and surveyors. The contribution of site management to estimating is to provide advice to the estimators on methods of construction and to discuss proposed method statements with the estimators.

(3)   External organisations

Material suppliers, plant hire companies and sub-contractors all become involved in the estimating process in that they receive and have to respond to enquiries for quotations from contractors. This also includes consultation with buyers and estimators.

THE ESTIMATING AND TENDERING PROCESS

The estimating and tendering process is presented in Fig. 2.1. This shows a logical flow of activities from the decision to tender to the tender submission. What a static flow chart cannot show adequately is the interactive nature of the whole process and the degree of interaction between the various personnel. These interactions vary from project to project and from estimator to estimator and to present one flow chart as 'typical' would be too bold; thus this chart is regarded as representative. It is based on information obtained from six companies.

The description of the process presented here is divided into three main parts: the decision to tender, the estimating process and the tender.

THE DECISION TO TENDER

The decision to tender is the starting point of the whole process. A contractor receives most of the opportunities to tender through being invited from select lists or by responding to advertisements. The invitation to tender or the details forwarded from advertisements may not comprise complete sets of contract documents. In these cases the client's advisors on behalf of the promoter are employing a form of pre-selection where brief details are issued and contractors can decide whether to proceed. The pre-selection information is likely to include:

- the names of the promoter and his advisors;
- the names of any consulting engineers with supervisory duties;
- the location of the site;
- a general description of the work involved;
- the approximate cost range of the project;
- details of any nominated sub-contractors for major items;
- the form of contract to be used;
- the procedure to be adopted in examining and correcting priced bill(s);
- whether the contract is to be under seal or under hand;
- the anticipated date for possession of the site;
- the period for completion of the works;
- the approximate date for the despatch of tender documents;
- the duration of the tender period;
- the period for which the tender is to remain open;
- anticipated value of liquidated damages (if any);
- details of bond requirements (if any);   and
- any particular conditions relating to the contract.

Based on this information the contractor's senior management is faced with the decision whether or not to request the contract documents and prepare a tender. This decision is based on such factors as:

- the company's current workload, turnover and recovery of overheads;
- the company's financial resources;
- the availability of resources to undertake the work;
- the type of work;

- the location of the contract;   and
- the identity of the promoter and his representative.

It is argued that the decision to submit a tender should result from the implementation of a company tendering policy from trading and marketing information such as:

(i)   turnover target, divided to show in which markets and in what proportions the total turnover can be obtained;

(ii)   overheads budget;

(iii)   gross and net profit targets;   and

(iv)   anticipated volume of enquiries required to achieve the turnover.

If a contractor decides on the basis of the pre-selection information to submit a tender and the client's advisors decide to invite that contractor, then a tender would normally be expected from the contractor.   However, the contractor's senior management may wish to review this decision after studying the contract documents.

If no pre-selection procedure has been adopted, as is often the case when using select lists of contractors, the contract documents are forwarded to the contractors invited together with a letter of invitation.   In these cases the contractor's senior management has the same decision to take based on the same factors except that the careful study of the contract documents can be undertaken at the same time.

THE ESTIMATING PROCESS

The process of producing a cost estimate on which a tender can be based is described in the following steps as represented in Fig. 2.1.

(1) programming the estimate;

(2) preliminary project study;

(3) material and sub-contract enquiries;

(4) project study, construction method and planning;

(5) calculating labour and plant costs;

(6) estimating the direct costs;

(7) calculating on-costs;

(8) preparing reports for the tender meeting.

(1) Programming the estimate

After the receipt of the contract documents and after the decision to tender has been taken, the estimator checks that all documents have been received. The estimator then prepares a programme of the tasks required to complete the estimate and establishes a list of key dates against which progress can be monitored. This is essential since the time to complete an estimate and tender is finite and the deadline for submission precisely defined. A major part of the management of the estimating process is to ensure that all tasks are given adequate attention within the limited time   available.

The estimator will also distribute copies of the documents to the relevant personnel, such as the planning staff.

(2) Preliminary project study

The next logical step in producing an estimate would be to undertake a complete and detailed project study from which a construction method would be selected and a pre-tender construction programme produced. However, the time required to obtain quotations from materials suppliers and sub-contractors is such that the enquiries have to be issued as early as possible. To enable this to be done a preliminary project appraisal is undertaken which establishes:

* the principal quantities of work;
* an approximate estimate;
* the items to be sub-contracted;
* the materials for which quotes are required;
* the key delivery dates;   and
* whether there is a case for considering design alternatives.

The preliminary project study is undertaken by both the planner and the estimator, if these are separate persons. The steps taken are:

(i)   to total all bill items containing similar classes of work. This establishes the principal quantities of work in each trade or class;

(ii) to use 'rough' or 'global' rates to cost or price these principal quantities and so establish an approximate value for the whole project and for each class of work;

(iii)   to create an outline or preliminary construction programme. This establishes the key delivery dates for the materials and sub-contractors;

(iv)   to list materials, quantities, specification and delivery dates;

(v)   to identify the items of work which are to be sub-contracted; and

(vi)   if the estimator considers that an alternative design would be cheaper, consultation with in-house designers would be started with a view to undertaking an outline design and calculating the new quantities.

Although the key dates and quantities may change as work on the estimate progresses, this preliminary study is essential to enable the major enquiries for materials and sub-contractors to be issued.

(3)   Material and sub-contract enquiries

(A)   *Obtaining quotations for materials price*

The   buyer's role

Due to the effects of inflation, the variance of delivery costs and quantity discounts, the estimator is usually required to obtain a quoted price for every major material to be included in the estimate. If not already done in the preliminary project study, this requires a detailed list of all the materials, their specification and the total quantities required to be abstracted from the contract documents. The delivery dates for the materials are obtained from the outline construction programme.

In most contracts the responsibility for sending out enquiries, chasing suppliers who are slow to respond and collating the quotations lies with the buying or purchasing department. In some companies this task is left with the estimator. Where a buyer is used to obtain quotations the instructions as to what is required are passed from the estimator. The quotations, once received, are passed back to the estimator for inclusion in the estimate. The usual function of the buying department in the production of an estimate is to provide a service to the estimator.

Fig. 2.1 The estimating and tendering process

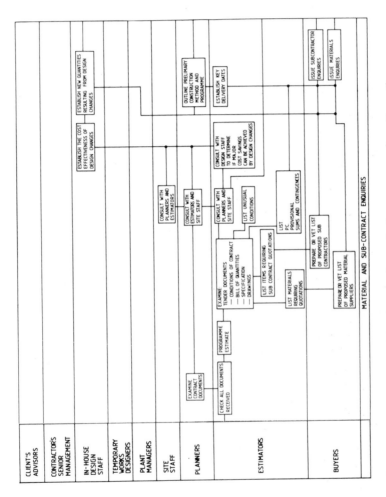

Fig. 2.1 The estimating and tendering process (continued)

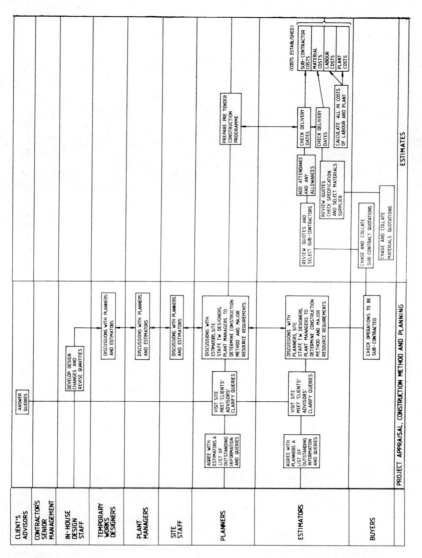

Fig. 2.1 The estimating and tendering process (continued)

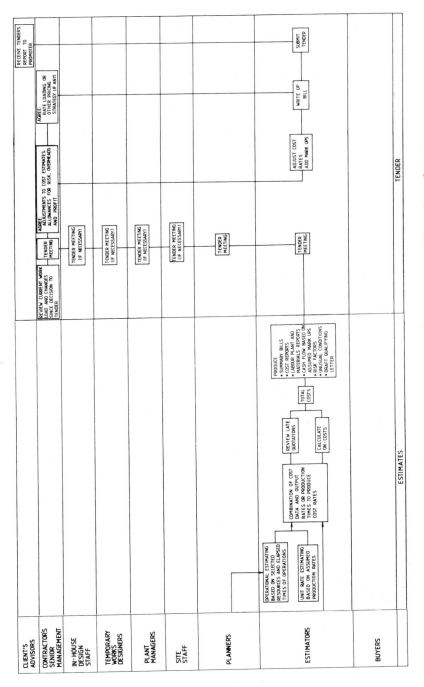

Fig. 2.1 The estimating and tendering process (continued)

Estimating in advance of quotations

Because of the constraint of time estimators are unable to wait for materials quotations to be received before proceeding with the calculation of costs for each item, or group of items of work.   It is, therefore, common practice amongst estimators to assume prices for materials and proceed with the cost calculations using these assumptions.   These assumed prices are based on recent quotations for other projects and the estimator's and buyer's knowledge of current price movements.   When the actual quotation is received the estimator's calculations are adjusted to take account of the difference between the assumed material price and the actual quotation.   If time permits this is done on an item by item basis.   If, as is frequently the case, time does not permit, a global adjustment will be made somewhere in the estimate to take account of these differences.

Enquiries

Enquiries sent to suppliers normally include:

- the specification of the material;
- the quantity of the material;
- the likely delivery programme including both the period for which supplies would be needed and the daily or weekly requirements;
- the address of the site;
- the means of access;
- any traffic restrictions and conditions affecting delivery;
- the period for which the quotation is required to remain either open for acceptance or firm;
- the date by which the quotation is to be submitted;   and
- the name of the person within the contractor's organisation to whom any reference concerning the enquiry should be made.

Buying departments usually have standard letters or forms for issuing enquiries and, where convenient, parts of the contract documents - such as extracts from the specification - are photocopied to accompany the enquiry.

As the estimator is responsible for completing the estimate an enquiry index or progress chart is usually kept so that the stage of each enquiry can be readily monitored.

Receiving quotations

The estimator is required when checking quotations to ensure that materials meet the specification of the contract documents and can be delivered to the site at the times required by the construction programme.   In addition the contractual obligations to be entered into for the supply of the material must be satisfactory.   The aspects of a materials quotation to be checked include that:

(i)   the quotation covers the actual material stated on the drawings;

(ii)   the material meets the standards described in the specification;

(iii)   the quantity is appropriate to the total quantity required in the works;

(iv)   the delivery period and programme meet the time required for incorporation in the works;

(v)   the discount rate (where applicable) is not less than the normal market rate;

(vi)   the trading conditions and terms of payment are acceptable;

(vii) the time limit which may be applied to the acceptance of the supplier's offer is acceptable;   and

(viii) whether the material is offered on a firm price basis or at prices ruling at the date of delivery.

The determination of the materials prices for inclusion in the direct cost estimate may be considered to be one of the most precise aspects of estimating. The process of obtaining materials prices as has been described can be seen to consist solely of contacting suppliers who have the material available and negotiating a suitable rate under satisfactory contractual conditions.   In addition the estimator has to undertake the more difficult task of determining allowances for material wastage, damage, theft and   delivery discrepancies in so far as they affect the costs of the works.

(B)   Obtaining sub-contractor quotations

The quotation and the estimate

As with materials quotations, sub-contractor quotations take time to assemble. Following receipt of the quotations they must be compared and the sub-contractors selected.   The rates for the selected sub-contractors will be included in the estimate together with allowances for attendance and other services. The profit to be added to the main contractor's own sub-contractors and to the nominated sub-contractors may be included at the estimating stage or left until the final additions following the tender meeting. Unlike materials quotations it is not common for estimators to make assumptions with regard to sub-contractors' rates but rather to wait until the actual quotations arrive before including them in the estimate. The difference between materials and sub-contractors is that in most cases the materials costs are combined with plant and labour to produce cost rates for items of work whereas the sub-contractors' rates in many cases will stand on their own together with an allowance for attendance. Thus major calculations are not delayed, as in the case of materials, whilst waiting for sub-contractor quotations.

As part of the preliminary study, the work that is to be sub-contracted would have been identified by the estimator. The factors that control the decision of which work to sub-contract are mainly the specialisation of the work involved and the size of the contract. Most contractors establish by practice the type of work they normally sub-contract. In undertaking contracts larger than usual a company may wish to sub-contract some work they would normally undertake themselves, the reason being to offset some of the financial risk. The absence of a direct financial risk in sub-contracted work is not a total security because of the indirect risk of losses caused by delay and disruption to the main works if the sub-contractors default. For this reason effective control of sub-contracted operations is important and this control begins with the selection of sub-contractors. Most companies keep a list of approved sub-contractors for various classes of work as a guide to estimators in the comparison of sub-contractor quotations.

Sub-contractors' enquiries

In issuing enquiries and negotiating a contract with a sub-contractor the buyer or the estimator should ensure that the following details are explicitly described and are agreed in writing, preferably as part of the quotation, but certainly as part of the contract that will be entered into between the main contractor and the sub-contractor should the work proceed.

(i)   The sub-contract programme should be stated in a reasonably detailed form, giving a realistic work sequence and a timed programme.

(ii)   The contract stage at which the service is required and the sub-contractor's commitments on either side of this date should be stated in order that unavoidable changes in schedule can be made as smoothly as possible.

(iii)   The sub-contractor's responsibilities with respect to making good other work should be stated.

(iv)   The position regarding the supply of equipment, access ways, storage facilities, etc. must be agreed.

(v)   Where the responsibility lies for the provision of water, power and any other services must be made clear.

(vi)   Specific instructions from the client regarding materials to be used or work practices adopted must be given.

(vii)   Facilities for inspections by the main contractor's representatives must be provided before the start of the contract.

(viii)   Responsibility for informing the appropriate body when work has to be inspected at various stages must be determined.

(ix)   Control information must be provided regularly by the sub-contractor while present on site.

(x)   Site safety and industrial relation requirements must be followed.

The above information is primarily obtained by the estimator from the contract documents and the construction programme. Appropriate sections of the conditions of contract and specification are usually copied direct from the contract documents and forwarded to the sub-contractor with copies of the drawings. A detailed abstract from the contract programme may be necessary to ensure the sub-contractor is fully aware of the periods he is required on site, the production levels needed and the interaction with other sub-contractors. As the sub-contractor enquiries are sent out as early as possible this will, in the first instance, be the outline or preliminary construction programme. As the construction programme becomes more fully developed a more detailed programme can be forwarded to the sub-contractors who are preparing quotations. Standard letters and sub-contractor enquiry forms are used to speed up the issue of information and to ensure that no basic contractual details are omitted.

As with materials enquiries a sub-contractor enquiry index is usually maintained in order to monitor the progress of each enquiry.

Receiving quotations

On receipt of sub-contractors' quotations the estimator is required to check that all the items have been priced correctly in accordance with the unit of measurement required.

The selection of the sub-contract price to be included in the estimate is not necessarily a question of finding the cheapest. Consideration is also given to all the factors known about the sub-contractor and any qualifications that may have accompanied the quotation. Where contractual problems exist they must be checked and agreed with the legal department.

Attendances

Having selected a sub-contractor the estimator does not simply enter the appropriate rate against the relevant items in the bill of quantities. An assessment has to be made of the attendance necessary for the sub-contractor and any particular site on-costs. Attendance falls into two categories:

(i)   General attendance or labours; and

(ii)   Special attendance or labours.

General attendance consists of:
-   Use of temporary works (e.g.. standing scaffolding);
-   Accommodation facilities, messrooms, sanitary and other welfare facilities;
-   Office accommodation and storage space for plant and materials;
-   Lighting, water power and other services;
-   Materials unloading and distribution; and
-   Clearing away of rubbish.

Special attendance consists of the provision of labours other than or in addition to those stated above, which are specifically described in the contract.

To calculate attendance requirements the estimator must fully understand the contract programme and the integration of the sub-contractors into the main works and have negotiated with the sub-contractors to determine fully their requirements on site. This may take the duration of the tender period and is carried on in parallel with building up the estimate.

(4)   Project study, construction method and planning

The project study is an activity that begins with receipt of contract documents and continues until the tender is submitted.   It is a process of continual refinement and revisions.   For the purposes of description it has been divided into two parts, the preliminary study already described and the main study which produces the construction method and pre-tender programme.   The key personnel are the estimator and the planner.   Where these are separate persons the exact division of duties varies from company to company and indeed in some companies one person undertakes all these tasks.   The objective is to choose the most efficient and hence cheapest construction method, to plan the construction around this method and to base the cost estimate on this method and plan.   To do this the planner and the estimator must gain a full appreciation of the work involved in the project.   This appreciation is gained by:

- A study of the contract drawings;
- An analysis of the bill of quantities and other contract documents;
- A site visit;   and
- The preparation of alternative methods of construction for the works.

In civil engineering contracts a copy of the contract drawings is sent to each company tendering for the work.   Any aspects of the project that are unclear from the details provided can be clarified by consultation with the client's advisors or their staff.

The analysis of the bill to produce principal quantities was described as part of the preliminary study.   This enables a broad understanding of the cost involved and an approximate value of each trade or class of work to be established.   The appraisal now being undertaken is more detailed, refining the initial analyses.

Site visit

It is normal practice for the estimator and planner to visit the site for construction of the works.   From this visit a report is prepared which gives details of:

- The description of the site;
- The positions of existing services;

- A description of ground conditions;
- An assessment of the availability of labour;
- Any problems relating to the security of the site;
- Details of access to the site;
- Topographical details of the site;
- A description of the facilities available for the disposal of the soil; and
- Details of any demolition or of any works and temporary works to adjoining buildings.

Preparation of method statements and planning

Method statements are descriptions of how the work will be executed with details of the type of labour and plant required and a pre-tender programme. It is in the preparation of these method statements that alternative methods of construction are considered together with alternative sequences of work, differing rates of construction and alternative site layouts. As these evaluations progress the preferred method of construction is chosen and the pre-tender construction programme illustrating this is prepared. In preparing the method statements the estimators and planners work closely and also consult with site staff, plant managers and temporary works designers and perhaps even consider re-designing part of the permanent works. The pre-tender programme prepared will show the sequence of all the main activities and their durations as well as the duration of the overall project. From this pre-tender programme approximations of the labour and plant resources will be calculated. The pre-tender programme is not an exercise carried out once only and left, but is subject to continual refinement and modification as both the estimator and planner become more and more aware of the implications of the project details. Thus throughout the preparation of the estimate the estimator and planner remain in close consultation. In many contracts the pre-tender programme is prepared in the form of a bar chart but in some of the larger and more complex projects some form of network analysis is used. Networks are also used in smaller contracts by companies who have become skilled in their preparation and use and are aware of their advantages. Bar charts and networks are described in Chapter 4.

It is in this part of the process of producing an estimate that the relationship between estimator and planner is at its most effective.

The amount of detailed information incorporated into the tender plan depends on the time available and the degree of competitive pricing considered necessary to obtain the contract. A secondary purpose of the tender plan is to provide a basis for the subsequent contract plan.

Design alternatives

If the initial study considered that an alternative design would produce a cheaper project, then the re-design process would have been initiated. If the main study confirmed this to be the case, then the design alternative would be worked up to a situation that produced outline drawings and quantities. The estimators would now be faced with adjusting the materials and sub-contractor enquiries and the estimate to take account of the new design. Estimates for both the project as described in the contract documents and the design alternative may be produced and two tenders submitted.

(5)   Calculating labour and plant costs

(A)   *Labour costs*

The all-in costs for labour are calculated by the estimator for each category of labour employed. An example of the calculation of labour costs is given in detail in Chapter 3.

Most civil engineering contractors limit the categories of labour to two, labourers and tradesmen; however, this is sometimes extended to three to include plant operators. It is interesting to note that building contractors tend to use many more categories and as many as fifteen have been observed.

(B)   *Plant costs*

The hourly or weekly cost of plant can be either as a result of internal calculation or from quotations. Methods of calculating hire rates are given in Chapter 3. Quotations for hire can either be internal rates from the plant department or the contractor's plant subsidiary, or they can be external hire rates from an independent plant hire company. Calculated rates or internal hire rates can be established very early in the estimating process. External hire rates may take a little longer but it is unusual to suffer serious delays in receiving quotations. Unlike materials obtaining plant costs is not regarded as a major problem. This is mainly because many of the data are obtained from within the contractor's own organisation.

(6)   Estimating the direct costs

The estimator's task is to determine the cost to the contractor of executing the work defined in the contract documents. This cost estimate will be modified by senior management in consultation with the estimator to determine the tender or selling price. The estimator establishes the direct cost rates for each item in the bill of quantities. A direct cost rate is a rate for the labour, plant, materials and sub-contractors, but exclusive of additions for site overheads, head office overheads and profit. These will be assessed and included later. Determining a direct cost rate involves selecting the appropriate resources of labour, plant and materials for the item of work (either a single bill item or a group of bill items), selecting the output or usage rates for each resource or determining the elapsed time that each resource, labour and plant, will be employed and combining this with the cost information collected. This combination of the unit cost of resources together with the usage of resources to produce a direct cost for the work described in the bill item or group of bill items is illustrated in Fig. 2.2.

The all-in rates for labour and plant are calculated at the start of the estimate. Materials and sub-contractor quotations are obtained during the tender period and incorporated as and when they are received. The production times or output rates used in calculating the direct costs are obtained by estimating techniques known as operational rate estimating, unit rate estimating, spot rates, and a number of other methods. These are described, with detailed illustrative examples in Chapter 3. Chapter 5 gives examples of the use of these estimating methods.

(7)   Calculating on-costs

When the direct cost for the project has been completed it is possible to assess the site overheads or on-costs. These may be summarised under the following headings:

(i)   site management and supervision;

(ii)   plant;

(iii)   transport;

(iv)   scaffolding;

(v)   miscellaneous labour;

(vi)   accommodation;

(vii)   temporary works and services;

(viii)   general items;

(ix)   commissioning and handover;   and

(x)   sundry requirements.

To aid the calculation of the sum of money required to cover these costs most companies have created   detailed check lists of requirements.   An example of a check list for supervision is given later in Table 5.50.

The direct cost estimate totals indicate the level of overhead support required for the project and the contract programme identifies the timing and duration of overhead requirements.

The check lists show the estimator what allowances should be made in each category. It is the estimator's responsibility to check that the latest rates for services, transport and site accommodation are included.   This involves communication with the company personnel department, plant department and external statutory authorities to ensure that all the requirements for the particular location are covered.

The calculation of site overheads is the last process in calculating the direct cost of the project. The estimator is then required to present to senior management reports relating to the project.   These reports contain:

- a brief description of the project;
- a description of the method of construction;
- notes of any unusual risks which are inherent in the project and which are not adequately covered by the conditions of contract or bills of quantities;
- any unresolved or contractual problems;
- an assessment of the state of the design process and the possible financial consequences thereof;
- notes of any major assumptions made in the preparation of the estimate;
- assessment of the profitability of the project;   and
- any pertinent information concerning market and industrial conditions.

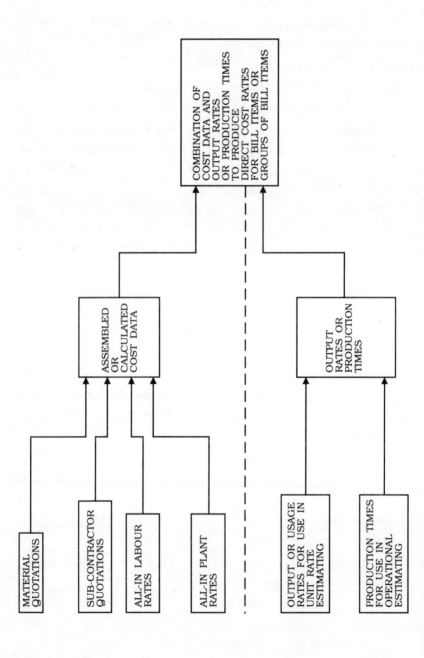

Fig. 2.2 Combination of unit costs with output or usage rates to produce a direct cost rate for an item

The costs of the work included in the estimate are reported to senior management in cost reports that give details of:

- main contractor's labour;
- main contractor's plant allocated to rates;
- main contractor's materials;
- main contractor's own sub-contractors;
- sums for nominated sub-contractors;
- sums for nominated suppliers;
- provisional sums and dayworks;
- contingencies;
- amounts included for attendance on domestic and nominated sub-contractors;
- amounts included for materials and sub-contract cash discounts;
- perhaps a bill of quantities marked up with the direct cost rates showing the labour, plant, materials, sub-contractor breakdowns for each rate.

As well as reporting the costs estimated for labour, plant and materials the estimators also assemble the total hours for each item of plant and total quantities for materials. These resource totals are compared with the planner's calculated resource totals and any differences reconciled.

The estimators may also calculate the cash flow for the contract based on a range of assumed mark-ups which will assist senior management's judgement as to what is the appropriate mark-up to select.

Examples of these reports are given in Chapter 6.

THE TENDERING PROCESS

The tendering process can be divided into four stages:

(1) the assessment of the estimate and evaluation of adjustments;
(2) the assessment of general overheads;
(3) the assessment of risk and profit allowances; and
(4) the writing up of the bill for submission.

(1)   Assessment of the estimate

Based on the reports prepared by the estimator the staff charged with the responsibility of submitting the tender will assess the estimate and decide what adjustments to make.   This group of staff will comprise senior management, the estimators, planners and other participants in the estimating process as required. It is the responsibility of this panel to satisfy themselves that the estimate is adequate. This is done by studying the reports prepared by the estimator and interrogating the estimator on his assumptions and decisions; on many occasions the estimate is adjusted, usually in the form of lump sum additions or subtractions.

(2)   Allowances for general overheads

A company's budget should include a forecast of turnover and a forecast of general overheads. This will establish the overhead allowance to be included in each tender. Monthly monitoring of a company's turnover and expenditure will give guidance as to what modification of this allowance should be made.

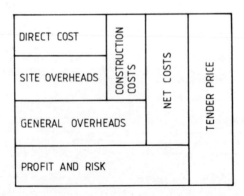

Fig. 2.3   Costs and additions in a tender sum

(3) Allowances for risk and profit

An addition for risk is included if the assessed chance of making a loss is greater than that of breaking even.

To assess the risk requires an assessment of the following information:

- the preliminary items within the bill;
- the work to be undertaken directly by the contractor;
- the total materials value and proportion of bulk materials;
- the work to be sub-contracted;
- the nominated sub-contractors ;
- the nominated suppliers;
- provisional sums and prime cost items;
- dayworks; and
- contingency allowances.

The profit allowance is usually assessed at what is considered to be possible in the prevailing market conditions.

(4) Writing up the bill

The structure of the costs and additions included in the tender price are as shown in Fig. 2.3. How these are presented in the bill is for the tender meeting to decide. Not every item need carry its 'true' share of costs and additions. If the rate for a bill item is increased then the rate for another must be decreased in order to maintain the overall total. Adjustments of this nature are known as 'rate loading'.

Rate loading is carried out when the total tender sum has been determined. This is the process whereby the rates for some bill items are raised while for others they are lowered, so keeping the total tender sum the same. The object of this 'loading' of bill items is:

(i) to make extra money on bill items;

(ii) to improve cash flow for the project; or

(iii) to make extra money through inflation indices.

(i) There may be items within the bill where the estimator considers that the quantity has been under-estimated. The opportunity might therefore exist in remeasurement contracts to make additional money by raising the rate of the item so as to capitalise on any increase in the actual amount of work completed on the project.

(ii)   By raising the rates of bill items relating to work completed early within the project the estimator increases the positive cash flow for the project and reduces the capital locked up in the project.

(iii)   During periods of high inflation when interest rates were less than inflation it was attractive to load the bill items relating to work to be completed in the later stages of the project. This enabled additional monies to be accrued following the calculations using the appropriate inflation indices. Due to Government policies these opportunities no longer exist.

The additions agreed by the tender meeting may be in the form of lump sum adjustments and percentages. Percentage additions may be made to either particular cost code categories (e.g. labour, plant, materials) or to individual bill items.

Lump sum adjustments are frequently made by using the preliminary section of the bill or the CESMM2 adjustment item to include the balance of the mark-up. The process of apportioning money throughout sections of the bill or particular classes of work is rarely performed because of the clerical effort required.

The transfer of the rates into the bill of quantities is carried out either manually by the estimator or by adjusting rates in the computer system. If a company is still using a manual process then the bill is passed to the comptometer operators for extension (the multiplication of each item rate by the quantity), totalling and checking. In computer systems this part of the process is largely automatic. The final stage is the returning of the completed bill of quantities to the promoter or his representative together with the form of tender formally offering to undertake the works.

Examples of the calculations made in the tendering process are given in Chapter 6.

3   The Calculations in the Estimating Process

INTRODUCTION

The calculations undertaken by estimators in the process of producing an estimate fall into two categories, calculating costs and calculating direct cost rates for items of work.

Fig. 2.2 showed the combination of costs per unit quantity of resource together with the quantities of those resources used (time in the case of labour and plant and quantities in the case of materials). This chapter deals with the principles of the calculations which determine the costs of labour, plant, materials and the rates for items of work.

The calculations explained are:

- 'all-in' labour rate;
- 'all-in' plant rate;
- cost of materials;
- direct cost rates:
  * unit rate estimating;
  * operational estimating;   and
  * other methods of estimating.

'ALL-IN' LABOUR RATE

To calculate the total hourly cost of labour to the company, the estimator must determine the basic hourly pay and all the numerous expenses resulting directly from the employment of labour. In civil engineering this hourly rate is usually calculated for three categories of workers - labourers, craftsmen and plant operators.

The main elements to be included in the calculations are:

- the basic or current wage rate inclusive of the cost of guaranteed minimum bonus payments;
- payments required by the Working Rule Agreement (WRA) (plus rates, tool money, dirt money, etc.);
- allowances to cover guaranteed time;
- sick pay allowances;
- employees' liability insurance;
- training levies;
- national insurance;
- pensions;
- allowances for holidays with pay;
- allowances to cover redundancy;   and
- wage related benefits.

The following annotated example shows the inclusion of all these factors.

To calculate the all-in rate the estimator must determine:

- the site working hours;
- the weeks worked in a year;
- the actual hours worked in the year;
- the non-productive overtime;
- the costs of employing labour;   and
- the annual cost of labour.

The information required for the calculations is obtained from:

- the construction industry standard agreements and statutory employment conditions (30), which give basic pay rates and allowances;
- the contract programme which gives the period of work;   and
- an examination of the site location and current market for labour, which gives guidance on normal hours worked and bonus levels.

The all-in rate calculated will vary for each contract and these differences will arise from the different allowances made by the estimators.

SITE WORKING HOURS

Assume the site works a summer and winter period. The actual hours worked per week will vary from   company to company. Assume a normal working week Monday to Friday to be 8.00 am to 5.30 pm in the summer period and 4.30 pm in the winter period with half an hour taken for lunch.

Summer period   - 30 weeks

Hours worked: Monday to Friday 8.00 am to 5.30 pm.

Monday to Thursday: 8 hours normal;   1 hour overtime

Friday: 7 hours normal;   2 hours overtime.

This gives a working week of 39 hours plus 6 hours overtime.

Overtime hours are paid at 1.5 times the normal rate.

Non-productive overtime   =   6.0 x 0.5 = 3 hours

Total paid hours per week =   39 normal hours + 6 overtime hours

                              + 3 non-productive overtime hours

                          =   48 hours

Winter period   -   22 weeks

Hours worked: Monday to Friday 8.00 am to 4.30 pm.

This gives a working week of 39 hours plus 1 hour overtime.

Non-productive overtime   =   0.5 hours

Total paid hours per week =   39 normal hours + 1 hour overtime

                              + 0.5 hours non-productive overtime

                          =   40.5 hours

The 'normal' working week is defined in the Working Rule Agreement (see Ref. 30) This provides for the normal working hours to be 8 hours per day Monday to Thursday and 7 hours on a Friday.   All other working periods paid at overtime rates.   In each working day a half hour is allowed for breaks which are not paid for.

TIME NOT WORKED DURING THE YEAR

Holidays

Winter   :   7 working days + Christmas, Boxing Day, New Year

                 Total   =   2 calendar weeks.

Summer   :   2 calender weeks.

Easter   :   4 working days plus Easter Monday.

Additional :   4 bank holidays during the summer period.

Total holidays   =   5 weeks 4 days.

Sick leave

Assume the time off work is 1 week in the summer period and 2 weeks in the winter period. Allowance is made in these 'all-in' rate calculations for sick pay on the basis of an additional 2% of the labour cost.

Inclement weather

Assume there are 60 hours lost due to inclement weather in the year.

THE ACTUAL HOURS WORKED PER YEAR

Summer period

45 hours are worked per week for a total of 30 weeks

=   45 x 30   =   1350 hours

Less

10 days summer holiday

4 days spring holiday

5 days public holiday

A total of 19 days of 9 hours   =   171 hours

Therefore the total hours worked in the summer period

$$= 1350 - 171 = 1179 \text{ hours}$$

Winter period

40 hours are worked per week for a total of 22 weeks

= 40 x 22 = 880 hours

Less

7 days winter holiday

3 days public holiday

A total of 10 days of 8 hours   =   80 hours

Therefore the total hours worked in the winter period

$$= 880 - 80 = 800 \text{ hours.}$$

Thus the total hours worked for a one year period

$$= 1179 + 800 = 1979 \text{ hours}$$

From this total a deduction must be made to make allowance for sickness and inclement weather.

Hours lost through sickness:

Summer period :   1 week of five 9-hour days

Winter period   :   2 weeks of five 8-hour days

Total hours lost   =   (5 x 9) + (10 x 8)   =   125 hours

Hours lost through inclement weather:

A total of 60 hours is assumed.

Therefore the actual hours worked in a one-year period

$$=   1979 - 125 - 60   =   1794 \text{ hours.}$$

DAY	M	T	W	T	F	S	S	TOTAL
Hours Worked	9	9	9	9	9	–	–	45
Hours to be paid at flat time rates	8	8	8	8	7	–	–	39
Hours to be paid at time and a half	1	1	1	1	2	–	–	6
Hours to be paid at double time	–	–	–	–	–	–	–	–
Total non productive time	0.5	0.5	0.5	0.5	1.0	–	–	3

Fig. 3.1   The calculation of non-productive overtime : Summer

DAY	M	T	W	T	F	S	S	TOTAL
Hours Worked	8	8	8	8	8	–	–	40
Hours to be paid at flat time rates	8	8	8	8	7	–	–	39
Hours to be paid at time and a half	–	–	–	–	1	–	–	1
Hours to be paid at double time	–	–	–	–	–	–	–	–
Total non productive time	–	–	–	–	0.5	–	–	0.5

Fig. 3.2   The calculation of non-productive overtime : Winter

CALCULATION OF NON-PRODUCTIVE OVERTIME

The calculation of costs relating to non-productive overtime is confined to the overtime which is worked as normal practice.   The cost of any additional overtime must be dealt with separately.

From the data in Figs 3.1 and 3.2 the total non-productive overtime may be calculated.

In the summer period there is a maximum non-productive overtime of 30 weeks x 3 hours = 90 hours. From this must be deducted the hours lost due to sickness and holidays. One week is assumed lost through sickness. Ten days are taken as summer holiday, four days as spring holiday and five days as public holidays.

The deductions are therefore:

Sickness	3	hours
Summer holiday	6	hours
Spring holiday	2.5	hours
Public holidays	3	hours
Total	14.5	hours

Therefore the non-productive overtime in summer

$$= \ 90 - 14.5 \ = \ 75.5 \text{ hours.}$$

In the winter period, the maximum non-productive overtime is 22 weeks x 0.5 hours = 11.0 hours. Two weeks are assumed lost through sickness, seven days through winter holiday and three days through public holidays.

The deductions are therefore:

Sickness	1.0 hours
Winter holiday	0.5 hours
Public holidays	Nil
Total	1.5 hours

Therefore the non-productive overtime in winter

$$= \ 11.0 - 1.5 \ = \ 9.5 \text{ hours.}$$

Total non-productive overtime for the year

$$= \ 75.5 + 9.5 \ = \ 85 \text{ hours}$$

COSTS OF EMPLOYING LABOUR

The employment of labour within the civil engineering industry in the UK is governed by the Working Rule Agreement produced by the Civil Engineering Construction Conciliation Board for Great Britain (30). This board, consisting of representatives of employers, trade unions and other trades bodies meets regularly to agree the basic pay rates and allowances for labour, plant operatives and craftsmen within the industry. The allowances that should be included in any calculation of the cost of labour are written into the rules of the Agreement.

When calculating the actual cost of employing labour consideration must be taken of all the following factors:

- the basic rate of pay as agreed under the current Working Rule Agreement Rule 1;
- any appropriate plus rate for the operation of plant or particular skills;
- overtime allowance;
- guaranteed minimum bonus;

- appropriate plus rates;
- holiday stamps and death benefit costs;
- sick pay;
- CITB levy;
- an allowance for severance pay;
- employer's and public liability insurance;
- small tool allowance; and
- an allowance to cover the cost to the contractor of the supervision of site labour.

Allowances may also be necessary in certain conditions for tool money, additional plus rates, and travel money. The Working Rule Agreement should also be studied for special conditions, night work, shift work, tide work, periodic travel, height money, etc.

There follows an example of the calculation of the total annual cost of labour and an 'all-in' labour rate.

CALCULATION OF THE ANNUAL COST OF LABOUR

The following example calculates the annual cost and the 'all-in' rate for a tradesman. Similar calculations should be produced for general labourers, plant operatives and other 'key' men as considered necessary by the estimator. The example includes typical allowances for bonus, tools, sick pay, etc. The current addition of the Working Rule Agreement should always be consulted to ensure that the latest appropriate allowances are included.

	£
Basic Rate x No. of hours worked	
283.5p x 1794	5085.990
Inclement weather	
Basic Rate x 60 hours	
283.5p x 60	170.100
Basic Total	£5256.090

Guaranteed minimum bonus	
£15.21 per 39 hour week for 47 weeks	714.870
Non-productive overtime	
Basic Rate x number of hours	

283.5p x 85	240.975
Public Holidays	
Basic Rate x 8 hours x 8 days	
283.5p x 8 x 8	181.440
Sick Pay Allowance (2% x Basic Total)	105.122
Plus rate of 2.5% on Basic Total to cover all	
Working Rule Agreement allowances	131.402
Paid Total	£6629.899

National Insurance @ 10.45% of paid total	692.824
Training levy @ 2% of paid total	132.598
Annual Holidays and Death Benefit	
£14.45 x 47 weeks	679.150
Small tool allowance and clothing allowance	
8% of the paid total sum	530.392
Travel allowance £25.00 per week x 47 weeks	1175.000
Supervision allowance	
6.5% of the paid total sum	430.943
Paid Total + Allowances	£10270.806

Allowance for severance pay	
2.5% of the paid total and allowances	256.770
Insurance	
3.5% of the paid total and allowances	359.478
Total annual cost	£10887.054

Calculation of the 'All-in' Rate

Total annual cost of craftsman	=	£10887.054
Total hours worked	=	1794
Therefore 'All-in' Rate	=	£10887.054/1794
	=	£6.069/hour

'ALL-IN' MECHANICAL PLANT RATE

The estimator is required to determine the total 'all-in' cost of each major plant item as an hourly or weekly rate. How this all-in cost is determined depends on the method of acquisition. The plant may be hired or bought. Hiring is the more common and construction companies either hire from their own plant division or subsidiary or from the specialist independent plant companies.

Hired plant

For hired plant the basic cost is the rate quoted by the plant company to which the estimator must add the cost of all additional running costs not covered by the hirer's quotation.   These running costs include:

* the emoluments of the operator;
* fuel;
* oil;
* grease;   and
* other consumables.

Most hired plant is provided with an operator but the estimator may have to allow for additional monies to bring the operator's rate up to that covered by any site agreement. Where no operator is provided, the full 'all-in' operator rate must be added.   Fuel costs are normally calculated on the basis of an estimated number of litres per hour for the plant.   To estimate this estimators use their own records and data from standard plant manuals.   Oil and grease are allowed for as an allowance per hour of plant time or a percentage addition to the fuel costs. Other consumables are allowed for by a percentage addition.

Bought plant

Where the plant item is owned by the contractor the rate used by the estimator must cover both the ownership and operating costs.   The following example shows the calculations for a crane.

To calculate the hourly rate for a crane the following information is required:

* Initial cost   £ 200935
* Resale value   £   14065
* Average working hours per year   2000
* Years of life of the machine   10
* Insurance premiums per year   £   1200
* Licences and tax per year   £   800
* Fuel at 20 litres per hour   £0.30 per litre

*   Oil and grease   10% of fuel cost
*   Repairs and maintenance   15% of initial cost
                                                    per year
*   Required rate of return on capital   15%

Overheads are not included for simplicity but are usually expressed as a percentage allowance on the hourly rate. These costs are included in the calculation of the hourly rate as follows:

COST ITEM	£ PER ANNUM
Depreciation:	
Based on a straight line over 10 years	
(200935 - 14065)/10	18687
Interest on finance, calculated using a capital recovery factor from interest tables (CRF = 0.199 @ 15% per annum for 10 years) where 200935 equals the initial cost of the crane = (200935 x 0.199 x 10 - 200935)/10	19892
Road tax and licences	800
Insurance premium	1200
Ownership cost £	40579
Consumables	
Fuel cost calculated on a consumption rate of 20 litres per hour for 2000 hours per year at £0.30 per litre.	
= 20 x 0.30 x 2000	12000
Oil @ 10% of the fuel cost	1200
Repairs 15% of £200935	30140
Operating cost £	43340

Total cost   = ownership cost + operating cost

= £405879 + 43340   =   £83919

Hourly cost   =   £83919/2000   =   £41.96/hour

The above calculation allows for the generation of sufficient income to replace the asset, cover operating costs and provide a return on the initial capital invested. If the company plant division is run as a profit centre and is expected to contribute to head office overheads and profit, then an additional allowance for profit and overheads must be added.

The hourly rate as calculated is that which the company plant department would produce for the estimator. An average number of hours per annum is used in the calculations. The estimator should assess if the hours the plant is used in the contract will vary from the average and make an allowance for extra fuel, oil and grease. Similar additions should be made for particularly tough working conditions. In some instances where the plant requirements are particularly heavy it may be decided to calculate the all-in plant rate excluding an allowance for fuel which is then treated as a separate resource within the calculations.

Whether using hired or internal plant the estimator works from a basic rate supplied or calculated and makes allowances according to the expected circumstances of the project. As with the 'all-in' labour rate, this rate is then used throughout the build-up of the direct cost estimate.

CALCULATING THE COST OF MATERIALS

Chapter 2 describes how quotations are obtained for the materials that are required within a project. The ultimate responsibility for obtaining the materials quotations rests with the estimator. Where the actual task is executed by a purchasing or buying department the estimator must communicate regularly with the personnel involved to ensure the correct information has been issued and the quotations duly received. When all the details of the quotation have been checked and the quotation accepted as satisfactory the price will be included in the calculations of the relevant work item.

The determination of the materials price for inclusion in the direct cost estimate is considered to be one of the most accurate aspects of estimating. The process of obtaining materials prices as has been described can be seen to consist solely of contacting suppliers who have the material available and negotiating a suitable rate under satisfactory contractual conditions.   The estimator must in addition undertake the more difficult task of determining material wastage, damage, theft and delivery discrepancies.   For some materials this may reach particularly high proportions and this aspect must be assessed by the estimator and reflected in the price included in the estimate.

The estimator has also to include an allowance for the off-loading and storage of materials as well as the other allowances.   Thus the material price used in the item build-up calculations may not be the quoted price but a higher price to allow for these variables.

An example of including such allowances is:

Quoted price of softwood for formwork	=	253.40 £/m³
Wastage allowance, 17.5%	=	44.34 £/m³

Allowance for off-loading, stacking and storing

= 0.125 man hours per m³ at £5.26 per man hour

= 0.125 x 5.26 = 0.66 £/m³·

Therefore the price of timber used in subsequent calculations

= 253.40 + 44.34 + 0.66 = £298.40/m³.

DIRECT COST ESTIMATES

Having calculated appropriate rates for the resources required to complete the project, the estimator must determine the quantity of each resource required for the various items of work in order that the total cost of completing the item may be calculated.

As described in Chapter 2 the estimator and the planner will engage in a detailed project study which will produce a suitable method statement and pre-tender programme.   The method statement and programme will detail the method of construction chosen, the sequence and duration of the key activities, the project duration and data on the labour, plant and material resources required.   During the project study alternative methods of construction will have been considered leading eventually to the selection of the preferred construction method.   The selection of this method will have been based on cost assessments and the alternative methods will have been compared in terms of total cost.

Within the framework of the preferred method statement the estimator has to calculate direct cost rates for each bill item or groups of bill items. The calculations involve the selection of resources required to execute the work and the determination of appropriate cost and production rates.   This is shown diagrammatically in Fig. 2.2.

The all-in rates for labour and plant are calculated at the start of the estimate.   Materials and sub-contractor quotations are obtained during the tender period and incorporated as and when they are received.

The output rates or production times used in calculating the direct cost may be obtained from unit rates or operational rates.   Unit rates are based upon output or usage rates, which may be:

* abstracted from previously recorded company manuals;

* taken from the estimators' 'personal' manuals;   or

* 'known' to the estimators by their experience.

All such output or usage rates are subject to modification by the estimator following an appraisal of the particular conditions of work under consideration.   Frequently such an appraisal is subjective.

The alternative to using such rates is to estimate the duration of the activities or operations involved and to base the calculations on the estimated costs of the labour and plant required for the duration of the work (as opposed to assuming any particular output rate).

These durations are derived:

* from the planning exercise where logic sequences determine the duration;   or

* by building up from several assumed output rates;   or

* by experienced judgement.

Calculations based on the duration of construction activities or operations are known as operational estimating.

Both operational and unit rate type estimating are used by civil engineering contractors' estimators, although neither method is adopted to the exclusion of the other.   Operational estimating is particularly prevalent in civil engineering estimating because such contracts have a large 'plant' element and plant is susceptible to idle time, which accrues costs that are not readily catered for in the unit rate approach.   Some contractors not only estimate the major plant dominated sections (e.g. earthworks) in this manner but also try to include as much of the remaining items of work in their operational estimating approach as is feasible to do.

If a truly operational estimating approach is adopted then the estimating process comprises planning, calculating the costs of the resources for each operation or activity and transferring these data to the bill of quantities. This approach is popular particularly with companies who have adopted network based planning techniques for most of the contracts. However, the major impediment to adopting this approach is the format of the bill of quantities.

The inclusion of method related charges to CESMM2 (5) encourages the concepts of operational estimating. However, the CESMM2 does not produce bills of operational format.

The detailed build-up for each bill item is calculated and recorded by the estimator on worksheets. A typical example of a worksheet is given in Fig. 3.3. The format of the estimator's worksheets may vary from company to company but the basic information on the item estimated (reference, description, quantity and unit) is recorded together with the cost and production rates. This information is extended to produce a total rate for the item that is split into the main cost code categories: labour, plant, materials, etc.

As and when the estimator considers each item and the information becomes available, the build-up is completed and the direct cost estimate total increases. Bill items are considered by the class of work to which they refer, e.g. all the excavation items together, all the concrete items together, and the estimator concentrates on each class of work in turn. It is important for the estimator to compare and adjust the rates for similar bill items to ensure that they correspond with one another. This may involve the item build-up being adjusted several times before finally being accepted by the estimator.

When accepted, the rate for a bill item is transferred to a photostat copy of the bill of quantities. It is normal for the estimator to subdivide the bill page into several columns to record the breakdown of the item rate. This enables the total of each cost category to be calculated for each page and hence the complete contract. An example is given in Fig. 3.4.

An example of unit rate estimating

In unit rate estimating the calculation of a labour, plant or material rate is based upon a predetermined output or usage rate and the quantity of work stated against the bill item. The calculation is usually performed on the estimating worksheets and then the rates transferred to the bill of quantities. An example follows.

Item	Description	Unit	Quantity	Rate	Amount
G526.0	High yield steel bars BS4449 Grade 460/425, diameter 20 mm, length 15 m	tonnes	6.350	£557.64	£3541.01

The estimator decides to employ the following resources:
* 20 mm diameter high yield rebar - cut, bent and delivered;
* Steel fixer;
* 22 R B crane (for off-loading the steel).

The output or usage rates, which are taken from company or personal data and amended to reflect the specific contract conditions, are:
* 20 mm diameter high yield rebar   1.10 tonnes/tonne
* Steel fixer   18.50 hours/tonne
* 22 R B crane   0.20 hours/tonne

A quotation is received for the supply of the steel and the all-in rate calculated for the steel fixer and 22 R B crane.

These are as follows:
* 20 mm diameter high yield rebar   £ 397.93/tonne
* Steel fixer   £6.30/hour
* 22 R B crane   £16.83 /hour

The direct cost rate for each resource is then calculated.

Resource rate   =   usage rate x cost

Rate for 20 mm diameter high yield rebar =   1.10 x 397.93

=   £ 437.72/tonne

Rate for steel fixer

=   18.50 x 6.30

=   £ 116.55/tonne

Rate for 22 R B crane

=   0.20 x 16.83

=   £   3.37/tonne

The rates are added to give the total item rate and the bill item amount is calculated by multiplying the rate by the quantity.

TOTAL ITEM RATE   =   £557.64

TOTAL ITEM COST   =   £557.64 x 6.35 = £3541.01

ESTIMATE SHEET

ITEM	MAN HOURS	DESCRIPTION	QNT	UNT	LAB	PM	TM	PLT	LABOUR RATE	PERM MAT RATE	TEMP MAT RATE	PLANT RATE
1/1/A		EXCAVATE AND CART AWAY.	2.95	M3								
	0.167	J.C.B.+ DRIVER @ 6M³/HR			7.27			19.30	0.56			1.67
	0.167	TIPPER.			7.27			8.00	0.56			0.75
									1.16			2.42
1/1/B	1.77	SUPPLY AND LAY 525MM DIA. CONCRETE PIPE	3.0	M		5.32 6.24			6.63	40.75		
	0.56	MASS. CONC. BED & HAUNCH	2.05	M3		543 40.0			8.53	1.96×0.397 11.51 60.26		

Fig. 3.3 An example of an estimator's worksheet

PART 3 – OUSTON LANE REALIGNMENT

Item No.	Code No.	Item Description	Unit	Quantity	Rate	Amount £	p
		LAB. PLT. MAT. **PIPEWORK – PIPES**					
1	13I3.1	500 mm Diameter standard concrete pipe in trench, maximum depth 1.4 m. 7.10 24.42 10.25	m	27	41.77		
2	13I3.2	150 mm Diameter porous concrete pipe in trench, maximum depth 1.4 m. 7.10 24.42 9.87	m	382	41.39		
3	16I2.1	75 mm Diameter PVC class D water main in 0.75 m deep trench, along verge in Ouston Lane. 1.92 1.08 3.12	m	425	6.12		
4	16I2.2	75 mm Diameter PVC Class D water main digging out old 1¼ " diameter water main in 0.75 m deep trench, on western side of Tadcaster Bypass. 2.95 3.14 3.12	m	50	9.21		
		PIPEWORK – FITTINGS AND VALVES					
5	J321	150mm Diameter concrete oblique angle junction 8.54 1.92 16.05	Nr	13	26.51		
6	J631	75 mm to 4" Diameter taper to connect Ouston Lane water main to existing 4" diameter water main. 10.31 2.52 18.23	Nr	2	31.06		
		PIPEWORK – MANHOLES AND PIPEWORK ANCILLARIES					
7	K252	Precast concrete catchpit with 1200 x 600 mm clear opening medium duty multiple triangular cover. 80.65 114.35 308.22	Nr	5	503.22		
8	K360	Precast concrete trapped gully 450 x 750 mm deep, with 450 x 500 mm clear opening grade D heavy duty double triangular gully grate and frame. 16.41 23.59 71.80	Nr	13	111.80		
					Page Total		

Fig. 3.4   The sub-divided direct cost rate

The total rate for each cost code (labour, plant, materials) is abstracted from the calculations.   In this example:

Labour	=	£116.55/tonne
Plant	=	£3.37/tonne
Material	=	£437.72/tonne

An example of operational estimating

Operational estimating is the calculation of a direct cost rate for labour and plant based upon the total quantity of work involved and the total period that resources will be required on site, that is for the elapsed time of the operation. This is illustrated in the following example.

Item	Description	Unit	Quantity	Rate	Amount
F523.0	Placing of rein-forced concrete Grade 25P to bases thickness 300-500 mm	m³	12.71	£11.30	£143.62

An estimator is pricing the plant required to place concrete on a particular contract.   From the bill he knows that the total amount of concrete is 8000 m³.   He decides that concrete may be poured at an average rate of around 210 m³/week giving a total number of weeks for concreting at 38.   The maximum pour to the base slab is some 160 m³.   Concrete will be delivered 'ready mixed'.

He decides that the plant needed for the placing of the concrete will be:
* 2 No.   22 R B cranes;
* 4 No.   concrete skips;
* 6 No.   dumpers;
* 6 No.   vibrators.

The plant will be hired from the company plant division at a weekly rate, held on site for the full 38 weeks and used solely for the purpose of placing concrete.   The total plant cost for the operation is therefore:

Item	No.		Weekly Rate		No. Weeks		Cost
							£
22 R B crane	2	x	313.80	x	38	=	23848.80
Concrete skip	4	x	12.33	x	38	=	1874.16
Dumper	6	x	40.36	x	38	=	9202.08
Vibrators	6	x	33.40	x	38	=	7615.20
			TOTAL DIRECT PLANT COST			=	42540.24

$$\text{PLANT COST PER } m^3 = \text{£}42540.24/8000 = \text{£}5.317/m^3$$

A similar calculation will then be undertaken to calculate the labour cost of placing concrete per cubic metre. Assuming a concreting gang of one ganger and four labourers is required for the thirty-eight week period, the total labour cost for the operation is therefore:

Item	No.		Weekly Rate		No. Weeks		Cost
							£
Ganger	1	x	309	x	38	=	11742.00
Labour	4	x	238	x	38	=	36176.00
			TOTAL DIRECT LABOUR COST			=	47918.00

$$\text{LABOUR COST PER } m^3 = \text{£}47918.00/8000 = \text{£}5.99/m^3$$

Total labour and plant cost per cubic metre for the placing of the concrete = £5.317 + £5.99 = £11.307/m^3.

When the average cost of placing a cubic metre has been calculated the rate may be applied to all the bill items that make up the total quantity of 8000 m^3 involved in the calculation.

In the case of some bill items the estimator may combine a unit rate and an operational estimating rate as in the following example, where the provision of concrete is on a unit rate basis and the placing involving labour and plant is on an operational basis.

Consider the following bill item:

Item	Description	Unit	Quantity	Rate	Amount
A	In-situ concrete Grade 25P to bases thicknes 300-500 mm	m^3	56.0	£64.54	£3614.24

The placing of the concrete to the bases was calculated on an operational estimating basis, as in the previous example, giving a rate of £11.30 /m³. The provision of concrete is calculated on a unit rate basis. Suppose the concrete is to be supplied by ready mix at £48.40/m³. The estimator allows 10% wastage so the rate for material = £48.40 x 1.1 = £53.24/m³. This gives a total rate for the item of £53.24 + £11.30 which is £64.54. The total item amount is £64.54 x 56 which is £3614.24.

THE SMALLER ITEMS

A bill of quantities may contain several thousand items. If items are ranked largest to smallest and plotted against the cumulative value of the bill, then it may be seen that a large proportion of the total value of the contract is contained within relatively few bill items. This is illustrated in Fig. 3.5 and shows that, in this case, 80% of the value is represented in 20% of the items. These proportions vary from contract to contract but the general pattern is consistent.

This distribution of contract value is important to the estimator in both the attention given to the bill item and the method of estimating. The format of the bill of quantities requires the estimator to enter a rate against every bill item. Many of the bill items are not very significant to the total value of the bill and are, therefore, estimated as quickly as possible using methods that do not involve a detailed assessment of the resource requirements to complete the item of work. These methods of estimating include:

* Spot rates;
* Included in;
* Item sum;
* Sub-contractor rates; and
* Provisional sums and prime costs.

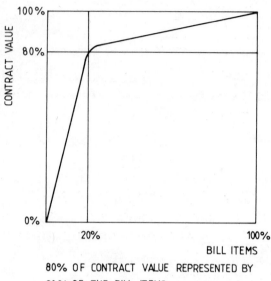

80% OF CONTRACT VALUE REPRESENTED BY
20% OF THE BILL ITEMS

Fig.   3.5   The distribution of bill items by value

Spot rates

A method for the less significant items is to estimate the cost rates for each
cost code category (labour, plant, materials, etc.).   This method is known as
estimating by 'spot' (or 'gash') rates.   The rates for each cost category are
included in the item rate based upon the estimator's previous experience.   The
total rate for the items is calculated by the addition of the rates for each
separate cost code.   An example is given as follows.

Item	Description	Unit	Quantity	Rate	Amount
C	Break out existing mass concrete	m^3	1.5	£62.90	£94.35

It is decided not to estimate the cost of this item by consideration of
the resources required but to price the item on a spot rate basis.   The
estimator simply enters the appropriate rates directly on to his work sheet:

* Labour   £42.50/m^3
* Plant   £20.40/m^3

Total rate   =   42.50 + 20.40   =   £62.90/m^3

Total item cost   =   62.90 x   1.5   =   £94.35

Included in

The estimator may simply mark a bill item as 'included-in' another. This indicates that the work involved has been considered within the general content of the project but that it is not considered appropriate to enter a rate against this item separately.

Item sum

A bill of quantities may contain items for which no quantity is given, the estimator being required to enter a single sum of money to cover the work involved. This method is normally used to price general items covering contractual requirements and specified requirements such as the testing of materials and the provision of temporary works.

SUB-CONTRACTOR RATES

The estimator has to decide at an early stage within the time allowed for the preparation of the estimate which items of work within the project are to be sub-contracted. Even though work may be sub-contracted it is important that the estimator has a reasonable understanding of what is involved in order to have sensible discussions with the sub-contractors. The estimator should calculate approximate rates independently in order to have a check on sub-contractor quotations. The process of obtaining and validating sub-contractor quotations is described in detail in Chapter 2.

When it has been decided to use a sub-contractor it is essential that the bill rates entered by the estimator are compatible with the sub-contractor's rates.

PROVISIONAL SUMS AND PRIME COSTS

Some of the items within the bill of quantities include within their detailed description information from the client's advisors as to how they must be priced.

Included within this category are:
*   Provisional sums;
*   Prime cost sums.

An abstract from a bill of quantities showing typical examples of these items is given in Fig. 3.6.

Provisional sums

A provisional sum item provides for a provisional sum to be included in the bill of quantities for the execution of work or the supply of goods, materials or services or for contingencies. At the time the tender documents were issued these requirements could not be entirely foreseen, defined or detailed. The sums may be used in whole or in part or not at all at the discretion of the client's advisors. They remain unaltered by the estimator and as such do not carry any profit. In tendering for work where such sums are included in the bill the element of profit on these items must be allowed for elsewhere.

Prime cost sums

A prime cost item is an item which provides for work or services to be performed by a nominated sub-contractor, statutory authority or public undertaking or for goods to be obtained from a nominated supplier. In the case of the nominated sub-contractor the client's advisors use this facility to ensure that a particular specialist sub-contractor performs the work specified. A nominated supplier of materials or goods is chosen by the client's advisors to supply particular goods or materials.

The prime cost sum entered against the item by the client's advisors is an estimate of the cost of the work to be performed by the nominated sub-contractor. Each prime cost item given in the bill of quantities must be followed by:

* an item, allowing the contractor to insert a sum for labours in connection with the prime cost item; and
* an item, providing for the contractor to insert as a percentage of the prime cost item all other charges and profit.

Number	Item Description	Unit	Quantity	Rate	
				£	p
A411	GENERAL ITEMS (contd) Provisional sums Daywork Labour	sum		5000	00
A412	Percentage adjustment to Provisional Sum for daywork labour	%			
A413	Materials	sum		2800	00
A414	Percentage adjustment to Provisional Sum for daywork materials	%			
A415	Plant	sum		2800	00
A416	Percentage adjustment to Provisional Sum for daywork plant	%			
	Other Provisional Sums				
A420	Call charges Engineer's staff site telephones	sum		500	00
	Prime cost items which include work on site by a nominated sub-contractor				
A500.1	Air valves and installation	sum		1800	00
A500.2	Labours	sum			
A500.3	Special labours, as Specification clause 8.3	sum			
A500.4	Other charges and profit	%			
	Prime cost items which do not include work on site by a nominated sub-contractor				
A600.1	Screening chamber screens	sum		600	00
A600.2	Labours	sum			
A600.3	Other charges and profit	%			
	to Part 1 Summary		Total		

Fig.   3.6   Examples of provision sums and prime cost items

4 Planning and Estimating

WHO PLANS FOR THE ESTIMATOR

The preparation of an estimate for a civil engineering project is heavily influenced or dominated by planning and the preparation of a construction method and the pre-tender construction programme. The reason for this strong link with planning is to be found in the nature of civil engineering work. Most civil engineering projects use heavy construction plant extensively and the preferred estimating method for plant dominated work is *operational* estimating. As described in Chapter 3 operational estimating is based on the selection of labour and plant resources and an estimate of the elapsed time of the operations on which these resources will be deployed. This estimate of an operation's duration is obtained from the construction programme. The existence of a pre-tender construction programme is thus essential to the estimating of the major work items in civil engineering projects.

In CESMM2 bills of quantities the opportunity to use 'method related charges' suits an operational estimating approach. In the remainder of the CESMM2 bill and in other bills the planners and/or estimators have to undertake the grouping of items into 'operations'.

As well as a timetabled expression of the construction method which establishes the sequence and duration of the key operations the pre-tender programme is also required to establish the sequence and delivery dates of materials, the dates each specialised sub-contractor is required and the duration of the on-cost services such as supervision, transport, welfare, offices, etc.

The preparation of the pre-tender programme is undertaken either by the estimator or by a planner. Both approaches are in current practice. The most common approach is to have both a planner and an estimator working in close liaison. The exact division of responsibility between these two persons varies between companies and between personnel within the same company. Chapter 2 describes the process of producing an outline pre-tender programme early in the estimating process and refining this into a construction method statement and a more carefully considered pre-tender programme as the project study continues. The preparation of these programmes for the chosen construction method and it's alternatives are mainly the responsibility of the planner. The evolution of the construction methods and the debate on the alternatives are a burden shared between estimators and planners with contributions from site staff, plant managers and temporary works designers.

PLANNING TECHNIQUES

The planning techniques employed in the preparation of the pre-tender programme are:

* Bar charts;   and
* Network analysis.

Of these two techniques bar charts are pre-eminent because of their simplicity. Companies who have become skilled in the use of network planning techniques and aware of their advantages use a network based approach to create the 'plan' and to explore the alternatives but the results are almost always presented in the form of a bar chart.

BAR CHARTS

Bar charts can be presented in two forms:

(a) simple bar charts;   or
(b) linked bar charts.

(a)   Simple bar charts

Fig. 4.1 shows a bar chart. The typical features of a bar chart are a list of activities on the left; a time scale marked out running from left to right; and the start, duration and finish of each activity represented by a bar. The level of detail of a pre-tender programme is usually cruder than that which will eventually be used for construction on site. For example, an activity in the

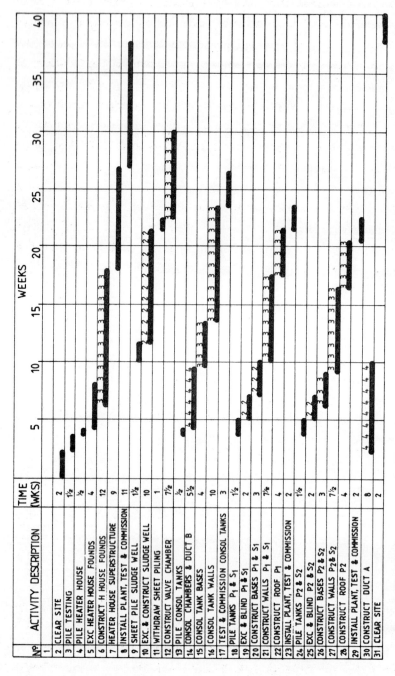

Fig. 4.1 A bar chart

(NUMBERS SHOW THE LABOURERS FOR CONCRETE AND EXCAVATION REQUIRED FOR EACH ACTIVITY)

pre-tender programme may be as broad as 'Construct Foundations' whereas the construction programme may divide this activity into finer levels of detail such as 'Excavate'; 'Blind'; 'Fix Reinforcement'; 'Erect Side Shutters'; 'Place concrete'; 'Cure'; and 'Back Fill'.   The time units used in a pre-tender programme are almost always weeks whereas some construction programmes will use days.

To create the bar chart the planner must list the activities, decide the start and duration of each and, implicitly, the interrelationship between the activities.   The difficulty in producing a bar chart is to take all these steps in one.   One advantage of network analysis is that the creation of a plan is broken down into smaller steps.

The planner can use the bar chart to create resource aggregation charts, as shown in Fig. 4.2.   This is achieved by allocating the resources (labour, plant or materials) to each of the bars and aggregating them to present a resource usage profile over time.

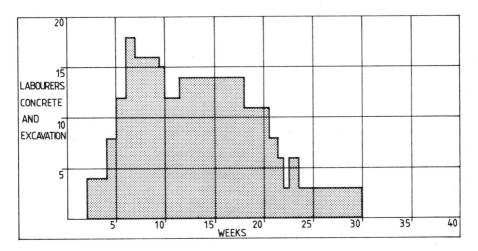

Fig.   4.2   A resource aggregation chart

Bar charts can also be used to explore the effects of different sequences of activities on the resource usage profiles and so determine the most efficient construction method and sequence.   It is in exploring alternative sequences of activities and the re-scheduling of activities to produce different resource profiles that the main limitation of the simple bar chart becomes apparent. This limitation is that the interrelationship between activities is not explicit;

efforts to overcome this have led to the use of linked bar charts.

(b) Linked bar charts

Fig. 4.3 shows the bar chart from Fig. 4.1 presented as a linked bar chart. The differences are that the links or the dependencies between the activities are illustrated by the inclusion of the vertical 'lines'. This now makes clear which activities must be complete before another can start. For example, it shows that activity Pile Tanks P1 and S1 cannot start until the activities Clear Site and Pile Testing are completed. It also shows that activities Pile Heater House, Pile Consol Tanks, and Pile Tanks P2 and S2 are waiting on the completion of the same activities Clear Site and Pile Testing.

This illustration of dependency also distinguishes between activities which have some 'extra time' or 'float' available before any delay interferes with the start of another activity and activities that have no float. The completion on time of these activities with no float is 'critical' to the start of subsequent activities and the completion date of the project. Activity Construct Duct A is an example of an activity with float and Fig. 4.3. shows that this activity has 28 spare weeks before a delay in its completion would delay the start of the succeeding activity Final Site Clearance. Activity Heater House Superstructure is an example of a critical activity. Any delay in the completion of this activity would interfere with the start on activity Install Plant and delay the completion of the project. There will be at least one continuous chain of critical activities running from the start of the project to the end. This chain is referred to as the critical path and is the chain or sequence of activities that controls the overall duration of the project.

By extending the simple bar chart to a linked bar chart the implications of manipulating the sequence of activities to alter the resource usage profiles become more obvious. These linked bar charts are more popular with planners and estimators than simple bar charts.

The weaknesses of a bar chart approach are:

(i)   the creation of the original bar chart is not broken down into small steps and this creates difficulties, particularly in complex projects; and

(ii) if the planner wishes to use computer programs to aid the analyses then the plan has to be in 'network form' to make use of the commercially available software.

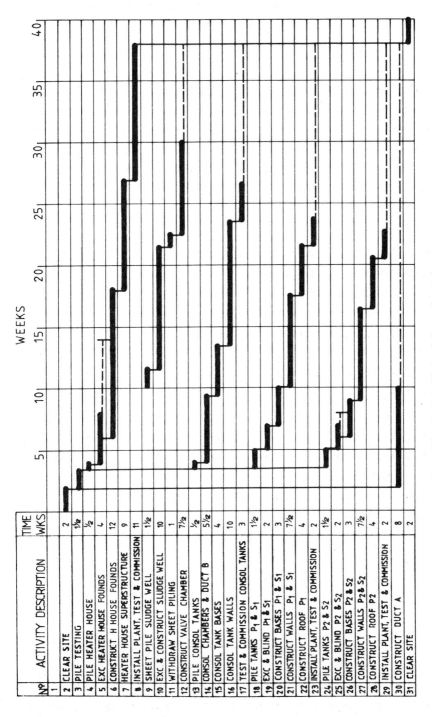

Fig. 4.3 A linked bar chart

For both these reasons it is now fairly common to find pre-tender programmes prepared from network plans with the aid of computers.

NETWORK ANALYSIS

Network planning in the end offers no more than linked bar charts. The advantages are that the smaller self-contained steps used in creating the network and experimenting with the sequence of activities are more applicable to complex projects and that computers can be employed to aid the analyses. It is also argued that the greater rigour imposed in preparing the logic diagram produces more realistic models of the proposed sequence of work. The steps in producing a network are:

(a) listing the activities;

(b) producing a logic diagram;

(c) estimating the duration of each activity and calculating the start and finish times of each activity and the float available; and

(d) estimating the resources required by each activity and

    (i) calculating resource aggregations,   or

    (ii) undertaking resource allocation exercises.

In producing a bar chart steps (b) and (c) are taken in one.

The two approaches to network analysis are activity on the arrow and activity on the node (precedence diagrams).   What evidence is available suggests that both approaches enjoy similar popularity in use for pre-tender planning.

Activity on the arrow

The preparation of a network plan based on activity on the arrow is as follows:

(a) Listing the activities

A list of activities is created and as for a bar chart the level of detail is commensurate with the planner's needs.

(b)   Logic diagram

In activity on the arrow systems the activity is represented by an 'arrow'. As the initial logic diagrams are not normally drawn to a time scale the length of each arrow has no significance.

The arrows, each representing an activity, are joined together in a logic relationship. As each arrow is drawn into the network the planner should ask three questions in order to check that correct logic is maintained.   The three questions are:

(1)   Which activities must be complete before this activity starts?

(2)   Which activities cannot start until this activity is complete?

(3)   Which activities have no logical relationship with this activity and therefore can take place at the same time?

If the constraints of resources (labour and plant) are ignored then the diagram produced will be one that shows the logical relationship between activities. Fig. 4.4 gives an example of a logic diagram for the same project represented by the bar charts in Figs 4.1 and 4.3.

In producing this logic diagram the 'purists' would argue that resources are ignored. However, in practice on site, planners planning construction operations tend to know what resources, particularly major plant items, are available and incorporate this within the production of the logic diagrams. That is, the planners build in resource constraints to the logic diagram. Planners producing a pre-tender programme are more likely to adopt a purer approach but are also known to build in resource constraints if these are known to them.

The junctions of arrows are called EVENTS and it is by numbering the events that the activities are identified. For example, from Fig. 4.4 activity Heater House Superstructure would be identified as activity 7-8.

Dummy arrows are used to introduce an extra event when two activities have the same start event and the same end event, otherwise unique identification of the two activities would not be maintained. Fig. 4.5 illustrates this. Dummy arrows are also used to maintain correct logic as Fig. 4.6 shows. Without a dummy arrow the start of activity D would be shown to depend on activities A and C and not just on activity C as shown.

(c)   Durations and time analysis

The duration of each activity is estimated and marked alongside each arrow as shown. The time unit used in pre-tender planning is usually weeks. The planner will obtain the estimates of duration from his own knowledge of the work, experience, historical records, work study data and discussions with estimators and site staff. If the durations of individual activities can be agreed then the duration of the overall project or sections of the project is unlikely to be disputed.

The earliest possible time of each event is calculated and written in the left hand square alongside each event. The calculations in Fig. 4.4 are event 1: earliest time is zero; event 2: earliest time is $0 + 2 = 2$; event 3: earliest time is $2 + 1.5 = 3.5$ and so on.

The point to watch is where two paths of activities merge, as for example at event 7; in all cases the longest path determines the earliest possible time of the event. Approaching event 7 from event 6 would give a time of $8 + 4 = 12$, whereas approaching event 7 from event 5 gives a time of $6 + 12 = 18$; as the longest path running into event 7 this determines the earliest time for event 7.

When the earliest time of each event has been calculated for all events the latest possible time of each event is calculated, starting with the last event and working backwards. The latest time of each event is written in the right-hand square alongside each event. The calculations in Fig. 4.4 are event 38: latest time is 40; event 9: latest time is $40 - 2 = 38$; event 8: latest time is $38 - 11 = 27$ and so on. Again the point to watch is when two paths leave the same event, as for example event 5 in Fig. 4.4; in all cases the earliest of the calculated times is taken. For example, for event 5 the latest time coming from event 6 would be $14 - 2 = 12$ and the latest time coming from event 7 would be $18 - 12 = 6$. In this case 6 is taken as the latest possible time; if a time later than this were taken, the chain of activities 5-7, 7-8, 8-9 and 9-38 would not be complete by week 40, the project end date. From these calculations of event times the start and finish times of each activity can be extracted. It should be noted that the times calculated are 'event' times and not 'activity' times. The event at the 'tail' of an arrow is the 'start' event and the event at the 'head' of an arrow is the finish event for that activity. From Fig. 4.4, for activity 30-31 event 30 is the start event and event 31 is the finish event. The duration of activity 30-31 is 1 week. The times of the start event define the limit of interference from the preceding activity and the times of the end event define the limits of interference of this activity on the

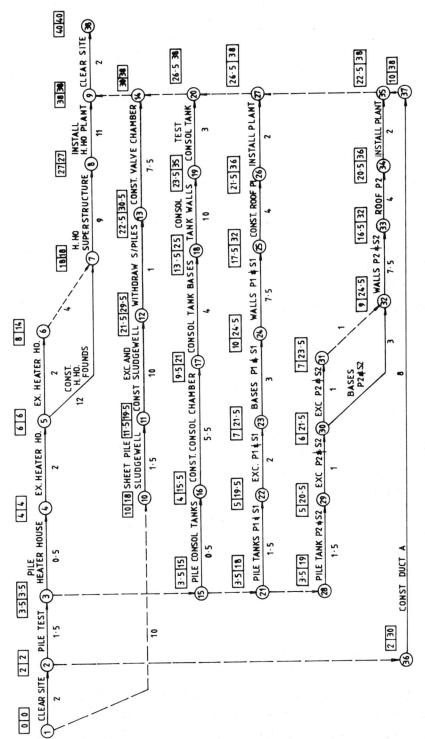

Fig. 4.4 An arrow network

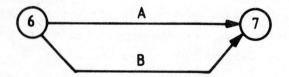

ACTIVITIES A AND B NOT UNIQUELY IDENTIFIED

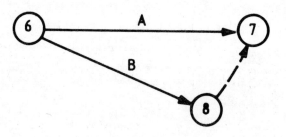

ACTIVITIES A AND B UNIQUELY IDENTIFIED

Fig. 4.5   The use of dummy arrows for unique identification

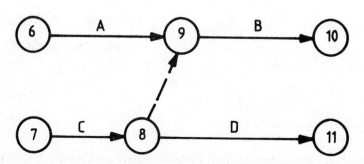

Fig. 4.6   The use of a dummy arrow to maintain correct logic

succeeding ones. The event times therefore define the time limits within which an activity may take place. The differences between event times and activity times for activity 30-31 from Fig. 4.4 are as follows:

Event Times		Activity Times	
Earliest time start event:	6	Earliest starting time:	6
Latest time start event:	21.5	Latest starting time:	22.5
			(23.5 - 1)
Earliest time finish event:	7	Earliest finish time:	7
			(6 + 1)
Latest time finish event:	23.5	Latest finish time:	23.5

The amount of 'spare' time or float each activity has can be calculated as illustrated in Fig. 4.7.

From Fig. 4.7 the 'total float' is calculated as the 'latest time of the finish event' less the 'earliest time of the start event' giving the 'total time available' less the 'duration'. The interfering float is the difference between the 'latest' and 'earliest' times of the finish event and is the amount of total float shared by the succeeding activity. 'Free float' is the difference between the total float and the interfering float. For the activity in Fig. 4.7 the calculated floats are:

Total float	=	74 - 45 - 10	=	19
Interfering float	=	74 - 66	=	8
Free float	=	19 - 8	=	11

For activity 11-12 from Fig. 4.4 the calculated floats are:

Total float	=	29.5 - 11.5 - 10	=	8
Interfering float	=	29.5 - 21.5	=	8

This indicates that all the total float is shared with the succeeding activities in the same chain.

The drawing of the logic diagram, the estimating of the durations and the calculations of the event times and float provide the data from which a linked bar chart can be drawn. Fig. 4.3 is the linked bar chart based on the network in Fig. 4.4.

(d) Resource assessments

Comments on the assessment of resources are included in the section on using computers for network analysis later in this chapter.

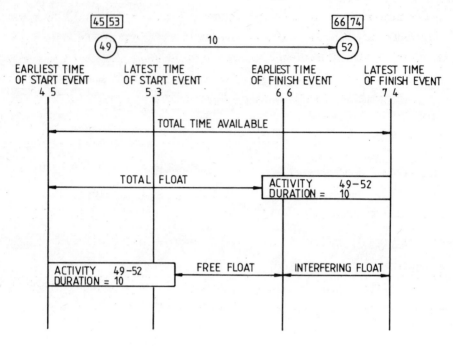

Fig. 4.7   Calculation of float

Precedence diagrams

The procedure for producing a precedence diagram follows the same steps as that for arrow networks.

(a)   Listing the activities

The activities are listed as before. This list may be extended by including the dependencies alongside each activity. Table 4.1 shows the activity list with dependencies. This shows that activity Clear Site depends on the start whereas activity Consol Tank Bases depends on the completion of activity Consol Tank Chambers & Duct B. Although it might be possible to create a list of activities and their dependencies without drawing the logic diagram it is more usual to add the dependencies to the list of activities as the logic diagram is thought out and drawn.

(b)   Logic diagram

In a precedence diagram or activity on the 'node' network the activity is represented by the node and the links or arrows are used only to show the logical relationships. In creating the logic diagram, as in Fig. 4.8, the same three questions as were asked in connection with 'arrow' networks are raised to ensure that correct logic is maintained, and the same comment on resources applies. Dummy activities are not needed to maintain correct logic or for unique numbering. The activities are identified by a single unique reference number.

Fig. 4.8 shows a precedence diagram for the same sequence of activities as was presented in arrows in Fig. 4.4.

Table 4.1   Dependency List

Activity No.	Activity Description	Depends On
2	Clear site	Start
3	Pile testing	2
4	Pile heater house	3
5	Exc heater house founds	4
6	Construct heater house founds	5 overlap 2 weeks
7	Heater house superstructure	6
8	Install plant, test & commission	7
9	Sheet pile sludge well	start + 10 weeks
10	Exc & construct sludge well	9
11	Withdraw sheet piling	10
12	Construct valve chamber	11
13	Pile consol tanks	3
14	Consol chambers & duct B	13
15	Consol tank bases	14
16	Consol tank walls	15
17	Test & Commission consol tanks	16
18	Pile tanks P1 & S1	3
19	Exc & blind P1 & S1	18
20	Construct bases P1 & S1	19
21	Construct walls P1 & S1	20
22	Construct roof P1	21
23	Install plant test & commission	22
24	Pile tanks P2 & S2	3

25	Exc & Blind P2 & S2	24
26	Construct bases P2 & S2	25 overlap 1 week
27	Construct walls P2 & S2	26
28	Construct roof P2	27
29	Install plant test & commission	28
30	Construct duct A	2
31	Clear site	8,12,17,23,29,30

(c) *Durations and time analysis*

The duration of each activity is estimated as before. The calculation of earliest and latest times for each activity proceeds in a forward pass to calculate the earliest times as before with the same rules for merging paths applying. Fig. 4.8 includes the calculated times. The major difference in this case is that the activities are represented by the 'node' and these calculated times are activity times.   The notion of event times does not exist and therefore the possibility of confusion is removed.   From Fig. 4.8 activity No. 10, Excavate and Construct Sludge Well, whose duration is 10, has the following activity times:

Earliest starting time of activity	11.5
Latest starting time of activity	19.5
Earliest finishing time of activity	21.5
Latest finishing time of activity	29.5

Activity No. 10 from Fig. 4.8 is the same activity as 11-12 from Fig. 4.4.

Float can be calculated in a similar way.   'Total float' is the 'latest finishing time' of the activity less the 'earliest starting time' of the activity giving the total time available less the duration.   'Interfering float' can only be calculated with reference to succeeding activities and is the 'latest finishing time' of the activity less the 'earliest starting time' of succeeding activities. 'Free float' is again the difference between 'total float' and 'interfering float'. For activity 10 from Fig. 4.8 the calculated floats are:

Total float	=	29.5	-	11.5 - 10	=	8
Interfering float	=	29.5	-	21.5	=	8
Free float	=	8	-	8	=	0

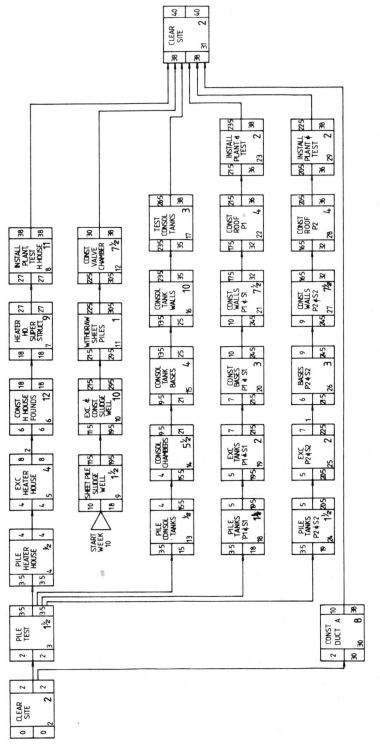

Fig. 4.8 A precedence diagram

From the drawing of the precedence diagram, the estimating of the durations and the calculation of the activity times and float the linked bar chart shown in Fig. 4.3 could be drawn. The arrow network in Fig. 4.4, the precedence diagram in Fig. 4.8 and the linked bar chart in Fig. 4.3 all represent the same sequence of activities.

(d)   Resource assessments

Comments on the assessment of resources are included in the next section on using computers for network analysis.

COMPUTERS AND NETWORK ANALYSIS

There are many commercially available computer programs which perform the calculations in network analysis. Programs are available for both activity on the arrow networks and activity on the node networks. The differences arise in the input, in particular the relationships between activities that are offered in the various programs.

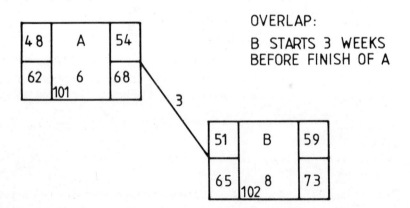

Fig. 4.9   Two overlapping activities

Arrow networks

The basic input data to arrow network computer packages required for each activity in the simplest programs are:

* activity identifier (e.g. start and finish event numbers);
* activity description;
* duration; and
* resource requirements for 'n' different resource types.

The general data input required by such programs which relate to the project as a whole will include data that prescribe the limits of each type of resource available to execute the project.

From the activity identifier the logic of the network is defined.   The identifiers for activities 30-32 and 31-32 are sufficient information to define the relationship between these two activities and activity 32-33.   It implies that activity 32-33 is dependent on the completion of activities 30-32 and 31-32.

On input of this information the available computer programs will:

* calculate event times;
* calculate float;
* produce tables which list all activities with start and finish dates, duration and float;
* draw bar charts with all activities starting as early as possible;
* perform time limited resource analyses and produce resource aggregation charts for all activities starting as early as possible and for all activities starting as late as possible; and
* perform resource limited analysis and re-schedule the sequence of activities to produce resource aggregation charts that are within the constraints of the prescribed resource limits.   Extract and list a schedule of revised activity start dates and draw a bar chart based on these scheduled start dates.

Precedence diagrams

The basic input data to precedence diagram computer packages required for the simplest programs are:

* activity identifier;
* activity dependency;

* activity description;
* duration;   and
* resource requirements for different resource types.

The general input data will again define the resource limits as with arrow packages. The difference between this and input for arrow networks is that in the latter the activity identifier was sufficient to define the logic of the network. In precedence diagrams the activity identifier is a single unique number and therefore is not sufficient for this purpose. Consequently an extra item of data is required to define the activity dependencies. For example activity 2 from Fig. 4.8 will be defined as depending on the start and activity 18 will be defined as depending on the completion of activity 3. Thus the dependencies that were included in the activity list when drawing the logic diagram are included in the input.

When the dependencies are included in the input as a separate item of data, it offers another degree of freedom in the definition of the relationship between the activities. In the computer programs for arrow packages the relationship between activities that was implicit was that of finish-start. That is activity 12-13 cannot start until activity 11-12 is finished. In the computer programs for precedence networks the requirement of defining the dependency separately allows a multiplicity of relationships to be defined. For example, Fig. 4.9 shows two activities whose relationship is not finish-start. This allows the overlapping of activities to take place in the final bar chart. The advantage of this is that in creating the original logic diagram the activities need not be broken down into smaller activities to achieve a logical representation as in arrow networks which do not have overlapping facility.

On input of this information the available computer programs will execute similar calculations and produce similar information to that described for arrow networks.

RESOURCE ANALYSIS

Fig. 4.2 shows the resource aggregation from the bar chart in the Fig. 4.1 chart for the resource of labourers. Within the limits of the float available (illustrated in Fig. 4.3, the linked bar chart), the starts of activities can be manipulated to smooth the resource profile removing 'peaks' and 'troughs'. Also the sequence of activities can be adjusted to accommodate a limited number of resource restrictions. For example the network may be drawn to take account of a construction method that uses, say, only one crane. This would mean that the activities requiring that crane would have to be done in series rather than in parallel. This style of resource analysis and manipulation of the activity sequence is representative of the type of analyses undertaken in the production of the pre-tender plan. More extensive analysis and manipulation usually require the use of computer programs. Resource analysis is described in detail in reference (1). With the use of computers resource analysis can be extensive; the forms of resource analysis are time limited and resource limited.

The input to computer programs for each activity describes the resource requirements (the number of each labour category and items of plant) for each activity and the general input for the project describes the resource limits. From this the various resource analyses required can be undertaken.

Time limited resource analysis

Time limited resource analysis is where the project end date is kept and any manipulation of the activities to alter resource usage profiles takes place within the float available. To aid this type of analysis, computer programs produce two resource aggregation charts for each resource being studied. The first resource aggregation charts are based on all activities starting as early as possible. This is similar to the resource aggregation chart shown in Fig. 4.2. The second resource aggregation chart is based on all activities starting as late as possible, that is each activity with float being delayed until all the float is used. These two resource aggregation charts represent two extremes. If some activities are delayed and not others, then other resource aggregation charts can be produced. Between the two extremes there is an infinite number of activity sequences and hence resource aggregation charts. The planner's task is to experiment between these extremes and find the 'smoothest' resource profile removing 'peaks' and 'troughs' in resource demands. This experimentation is usually done manually by visual inspection of the resource aggregation charts. In pre-tender planning the planners have a tendency to operate close to the situation where all activities start as early as possible.

Resource limited analysis

Resource limited analysis is where a specified resource limit is enforced and a sequence of activities is found whose resource demands do not exceed this limitation. The method of determining this sequence of activities is firstly to rank the activities into a priority queue. This means of deciding the priorities is by sorting the activities in some defined way. One of the common ways of sorting activities is to arrange them in order of 'earliest start': thus all the activities starting the first week would get their resources first and so on. As it is common to have activities with the same 'earliest start time' it is necessary to have a second sorting to distinguish these. A common second sort is 'total float'; that is, where two activities have the same 'earliest start time' the one with the least total float would be ranked above the one with the greatest total float in the priority list. There are several other sorts such as 'late start - total float', 'duration - total float', etc. Once the priority list has been created within the computer program a resource aggregation chart is drawn. The resource requirements of each activity are included in the chart in turn according to the priority list and if, when an activity's resource demand is included, the total resource demand is within the resource limit then the activity is allocated that starting date. Alternatively, when an activity's resources are added into the aggregation and the total resource demand exceeds the resource limit, the activity is delayed until it can be started at a time when the total resource demand stays within the limit. This highlights the importance of the priority listing as it determines the order in which an activity's resources are added into the aggregation charts and whether or not the activity starts at its earliest time or is delayed. In delaying an activity the computer programs are careful to check that the logic of the network is maintained and that activity 12-13 does not start before activity 11-12 is complete. If several, say four, resources are being examined simultaneously, then the logic of the network must be maintained across the four-resource aggregation charts and if activity 11-12 is delayed in one chart it is delayed in them all.

In resource-limited allocations the original project end date may be exceeded.

As the allocation process re-schedules the activities start date a new schedule of activity start dates is produced and usually a new bar chart based on these re-scheduled dates is drawn.

RELATIONSHIP BETWEEN ESTIMATING AND PLANNING

The relationship between estimating and planning is described in Chapter 2, in the section describing the 'project study' phase of the estimating process. It is illustrated in the flow chart in Fig. 2.2 and is referred to in the opening section of this chapter. Estimating and planning are not separate functions but part of the same process of producing an estimate and tender. The production of an estimate for a civil engineering project requires the production of a pre-tender plan. The creative process of deciding on the construction method and exploring alternatives tends to take place around the preparation of the pre-tender construction programme. The fact that estimating and planning tend to be described as separate functions is largely because the majority of companies have found it convenient to create a planning department as well as an estimating department. The planners often have duties other than producing pre-tender programmes. Their duties frequently include producing more detailed construction programmes for use on site and for providing planning services to site staff. The creation of construction methods, pre-tender programmes and estimates is in many cases too large a task to be undertaken by one person and the division of labour between planners and estimators has been a convenient division for companies to make. All companies with separate estimating and planning departments emphasise the close liaison that exists between the personnel. In the other cases the estimator is responsible for his own planning.

Barnes (15) explained that the influence of planning on the estimating process occurred in two main ways:

(i)   the two functions may be fully integrated, the contract programme being used as the basis for calculating the full cost and revenue characteristics of the contract. These figures are then transposed into the rates entered into the bill of quantities;

or

(ii) the two functions are not fully integrated and the planned use of the plant and labour resources is costed in total to be checked against the labour and plant cost totals as calculated by the estimators. The two functions are pursued in parallel with informal interaction limiting their divergence.

Situation (ii) is the more common.

The tasks where planning and estimating interact are as follows:

Task	Planner/Estimator
Study contract documents	Both
Produce outline tender programme	Planner
Calculate principal quantities	Estimator
Approximate estimate	Estimator
Identify items to be sub-contracted	Estimator
Establish which materials require quotations	Estimator
Key delivery dates	Planner
Decide if there is a case for design alternatives	Both
Obtain quotations for materials and sub-contractors	Estimator
Site visit	Both
Prepare method statement and alternatives	Both
Prepare pre-tender programme	Planner
Calculate labour/plant costs	Estimator
Estimate direct costs	Estimator
Estimate on-costs	Estimator
Reconcile planning and estimating data	Both
Prepare tender meeting reports	Estimator
Attend tender meeting	Both

RECONCILING PLANNING AND ESTIMATING

In most companies planners and estimators are employed and during the estimating process some of the work of the planners and estimators takes place in parallel with varying degrees of liaison between the personnel. Given that this is the case there arises the need to reconcile the calculations of planners and estimators. The reconciliation is a comparison to ensure that:

* the labour totals calculated by the planners and estimators are the same;
* the plant totals calculated by the planners and estimators are the same;

* the durations of key activities used by the estimators agree with the pre-tender programme;   and

* the duration of the project on which the estimator bases the calculation of site on-costs agrees with the pre-tender programme.

The amount of reconciliation between planning and estimating depends on the degree to which the two functions have diverged. If planning and estimating are undertaken by the same person then there should be no requirement for reconciliation. If the planning and estimating are undertaken by different persons the reconciliation required then depends on the closeness of the liaison between them and the methods of estimating employed by the estimator.

If the major items of work are estimated on an operational basis as described in Chapter 3, then the duration of each operation is taken from the pre-tender programme and no difference between estimators and planners should have been created. If the estimators and planners have agreed in determining the construction method, the resources of labour and plant to be deployed in the operations, then again no differences should be created. Operational estimating is the costing of the plan and reduces the need for reconciliation exercises.   Operational estimating is favoured by civil engineering estimators because of its suitability for estimating plant dominated work which includes idle time, down time and travel time. It links well with planning but requires manipulation to convert the calculated sums into bill item rates.   For this reason unit rate estimating is also used.

If the estimators use unit rate estimating as described in Chapter 3, where the calculations of the costs of items, or group of items, are based on assumed outputs of labour and plant with no reference to the duration of the operation, then differences between planners and estimators can occur.   For these items the estimator should convert the assumed outputs into elapsed times for the quantity of work and compare these elapsed times with the operation's duration in the pre-tender programme.   The difference between these is some in-built idle time.   If estimators are not careful, unit rate estimating which is based on an assumed average output of the plant may not reflect this idle item adequately. It is examples of this nature that cause divergences between the planned totals of plant and the estimator's calculated totals.

5 Examples of the Calculations within the Direct Cost Estimate

INTRODUCTION

This chapter contains examples of typical calculations to determine the direct cost of a construction project. The examples have not been taken from a single existing project but assembled from a number of different real contracts to produce a project that reflects a range of work. Examples are given for the following types of construction work:

* earthworks;
* drainage;
* formwork;
* reinforcement; and
* concrete.

A section is also included on site on-costs or overheads. Also included are the estimator's report on the project to the chief estimator, the site inspection report, a tender programme for the works and a method statement.

The output rates and costs used within the examples should not be taken as standards but simply as representative figures to indicate the methods of calculation employed by a construction estimator. The purpose of this chapter is to show how the construction estimator builds up the cost of bill items to form the overall project cost.

ESTIMATOR'S REPORT

On receipt of the contract documents the estimator must study them carefully and obtain as quickly as possible an understanding of the project, the specification for the works and the contractual conditions. The estimator is required to prepare an initial report outlining the project, and describing the works, their location and listing the main items of work and the quantities involved.

A typical (fictional) report is given below with the information separated under suitable headings:

TENDER NO:	89/115 - Oxbridge Bypass
LOCATION:	Oxbridge is a small Wessex market town situated on the main A362 trunk road approximately 14 miles west of Stourbridge.
THE WORKS:	The works consist of the construction of approximately 6.15 km of single carriageway road (7.3 m wide), three roundabouts and the diversion of several side roads. Associated drainage and accommodation works are included. The earthworks involve the excavation of some 279620 m^3 of material. There are three road overbridges, two culverts and a subway with associated retaining walls. The roadworks are of flexible pavement construction.

Bridge 1.   Downland Farm.   Span 19 m.
Reinforced concrete abutments.
Prestressed in situ concrete deck.
Bridge 2.   Pepperbox Farm.   Span 15 m.
Reinforced concrete abutments and deck.
Bridge 3.   Oakfield Farm.   Span 19 m.
Reinforced concrete abutments.
Prestressed concrete deck in M and U beams.

PRINCIPAL QUANTITIES:

Site Clearance:	20.06 ha		
Fencing:	Temporary 520 m.		
	Safety 1643 m.		
Earthworks:	Topsoil	28300	m^3
	Suitable for fill	234400	m^3
	Unsuitable for fill	6800	m^3

	Rock	1200	m³
	Dispose	5000	m³
	Imported fill	16100	m³
	Soiling	22980	m³
	Dispose topsoil	5320	m³
Drainage:	Pipelines 150 to 600 dia.	3600	lin m
	Ducts 100 dia.	400	lin m
	French drains 150 to 375 dia.	4800	lin m
	Manholes	75	No.
	Catchpit	38	No.
	Gullies	150	No.
Roadworks:	Sub-base	16000	m³
	Flexible carriageway	50000	m²
	Side roads	6373	m
	P.C. kerbs	10210	m
	Channels	200	m
	P.C. Edging	250	m
	Flexible footpaths	5000	m²
	Traffic signs	68	No.
Structures:	Excavation - suitable	7210	m³
	- unsuitable	1600	m³
	- rock	110	m³
	- dispose off site	1600	m³
	Imported fill	20170	m³
	Formwork	9480	m²
	Reinforcement	150	tonnes
	Concrete	2480	m³
	P.C. Beams	15	No.

SITE INSPECTION REPORT

As soon as possible the estimator will arrange a site visit to the location of the works.   This is necessary to obtain as much detailed information as possible about the physical nature of the site, the geographical surroundings and all available services and suppliers.

An example inspection report for the project is as follows:

TENDER: Oxbridge Bypass.

SITE VISIT BY: Stephen James and Simon Pearce.

DATE: 14th November.

WEATHER: Fine.

NAME AND POSITION OF PERSONS CONTACTED:

Mr J F Jones, Oxbridge County Council.

GENERAL DESCRIPTION OF SITE AND SURROUNDINGS:

GENERAL NEIGHBOURHOOD: (Rural, Urban, Wasteland):

Rural.

ADJACENT PROPERTIES:

(Protection and/or maintenance of access required).

Small housing estate at Oakwood Hill.

LAND AVAILABLE FOR OFFICES, STORES, PLANTYARD, CAR PARK:

Disused Railway Yard.   Availability must be checked with British Rail.

SPACE AVAILABLE AT STRUCTURES (Working area, traffic diversion, etc.):

Space available but extra land may have to be negotiated with farmers.

TREES, HEDGES, CROPS, LIVESTOCK:

Minimum site clearance required except for Spinney Wood.

Livestock adjacent to 80% of the proposed road.

NATURE OF GROUND:

(Low-lying areas, boggy patches, steep slopes, rivers):

Generally sloping but with two steep hills.   Well drained apart from area
adjacent to River Welford.

NATURE OF GROUND SURFACE AND SUITABILITY FOR VEHICLES:

Suitable for four-wheel drive and tracked vehicles.

EFFECTS OF BAD WEATHER:

Flooding to low level areas adjacent to river.

OVERHEAD CABLES:

11 KVA and 33 KVA as on drawings.

SPECIAL PROBLEMS;

(Tides, high winds, etc.):

River Welford flooding.   Road diversion at Oakley Hill.   Demolition of existing
bridge.   Access required to River Welford on each side.

ADJACENT PROPERTIES:

(Description and distance from site works):

Estate 15 m from site boundary at Oakwood Hill.

IS NOISE CONTROL NECESSARY?:

Not a contractual requirement but will be required in Oakwood Hill area.

ACCESS TO SITE:

(Possible routes, class of roads, restrictions on bridges, road widths, traffic density, etc.):

Two-lane bridges over River Welford and on Oakwood Hill Road both narrow. Existing A362 and the Oakwood Hill Road have high traffic density during peak summer periods.

ACCESS TO INDIVIDUAL STRUCTURES:

River Welford bridge north from Oxbridge District Council Car Park and south from Cranleigh Schell access road.

ROADS ACROSS SITE:

(Crossing points for site traffic, possible road diversions and closures):

Oakwood Hill, Ham Lane, Welford Road, Dale Farm, Cranleigh Drive, Park Avenue.

EXISTING FACILITIES:

(Buildings on site which could be used):

None.

SERVICES TO OR NEAR SITE WHICH COULD BE USED:

Water ?   -   Yes

Electricity?   -   Yes

Telephone?   -   Yes

Sewers?   -   Yes

All services at both Oakwood Hill and railway yard.

SITE CLEARANCE:

WILL TEMPORARY FENCING BE REQUIRED:

Some is included in the bill. Additional fencing may be required depending on the contract programme.

GENERAL AREAS, HEDGES AND TREES:

Cranleigh Wood and Oakwood Hill wooded area. Remaining areas are all hedge, fence and field.

SPECIFIC DEMOLITION ITEMS:

Existing railway bridge and farmhouse.

EXCAVATIONS AND EARTHWORKS:

OPEN EXCAVATIONS VISIBLE ON OR NEAR SITE:

Archaeological excavations at Dale Farm and Oakwood Hill Quarry.

NATURAL SLOPE OF GROUND:

Varies, see geological report.

NATURE OF EXPOSED ROCK:

Shales.

POSSIBLE TIP AREAS:

Railway cuttings.

POSSIBLE BORROW AREAS:

Embankments and Oakwood Hill area.

LOCATIONS OF PHOTOGRAPHS TAKEN:

None taken.

OTHER CONTRACTS IN THIS AREA. (The companies involved, size and nature of the work):

R. McDrew, Wadebridge Bypass, similar size.

LOCAL HAULAGE CONTRACTORS, QUARRIES, PLANT HIRERS, READY MIX PLANTS, ETC:

Haulage   -   Wilkinsons Haulage Ltd

Ready Mix -   Pioneer Concrete (Wadebridge) Ltd

Ready Mixed Concrete (Stockbridge) Ltd

A.R.C. (Stockbridge) Ltd

Plant Hire -   Reg Matthews Plant

Stockbridge Plant Co Ltd

Stevens Plant Hire

Wadebridge Plant Co Ltd.

THE CONTRACT PROGRAMME AND METHOD STATEMENT

The tender programme

An outline programme for the works is prepared. An example is given in Fig. 5.1.

A general method statement

The method for construction decided upon follows the normal pattern for the construction of motorways and trunk roads with the principal activity being the completion of the earthworks. The period is divided into two periods separated by the winter months when no earthmoving using scrapers would be possible due to the soil conditions.

Before the earthworks may commence it is important that the site is set up and all fencing erected to delineate clearly the limits of the site and prevent livestock adjacent to the road wandering onto the site.

All pre-earthworks activities such as:

*   'V' ditching to the top of embankments;

*   the diversion of existing services;

*   the construction of culverts across the line of the road;   and

*   the construction of temporary road diversions,

must be completed as soon as possible to give the earthmoving plant full access as required. Although programmed to finish in October the earthworks operations will continue for as long as possible before being halted.   The machines will be stored over the winter months when they will be serviced and overhauled.   Areas of 'cut' will be left with approximately 600 mm of excavation required as a protection layer to prevent the formation being damaged if during the first season of earthworks this level is approached. Drainage works will commence as soon as appropriate carriageway levels are approached and operations continued over the winter period. The landscaping to the earthworks will be completed as soon as convenient, not because the activity is critical for the programme but to ensure that embankment slopes are top-soiled and seeded prior to the rainy winter months that may cause damage to the slopes.

During the autumn and winter months work will concentrate on the bridges and other structures. These are carefully planned to allow the smooth movement of labour, plant and material resources from one structure to another. The completion of the subway structure is essential prior to the start of the second stage of earthworks.

The second stage of earthworks commences in the spring of 1990 and is due for completion in early autumn. The construction of the road follows the earthworks once a sufficient length of carriageway has been prepared to ensure a smooth flow of work. This continues into the early part of 1991. The 'black top' work will be sub-contracted to a suitable sub-contractor.

Detailed method statements

For each structure, detailed method statements will be prepared analysing each operation and the resources required. The use of critical path and other

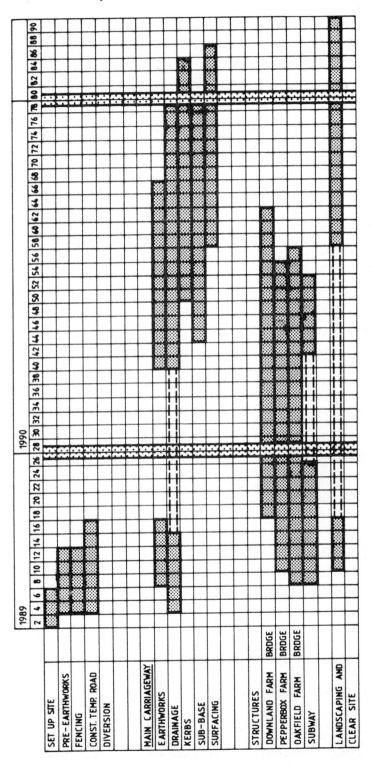

Fig. 5.1 A tender programme for the works

planning and scheduling methods may be incorporated to obtain as clear a picture as possible of the work to be undertaken and the problems involved. These programmes will be improved throughout the tender period to produce as detailed an analysis as time may allow.

LABOUR, PLANT AND MATERIAL RATES

Labour

The labour rates in Table 5.1 were calculated as indicated in Chapter 3.

From these rates the hourly rate for typical labour gangs may be calculated as ganger + 2 labourers =   £6.69 x 1 + £5.10 x 2 = £16.89 per gang hour or £8.45 per labourer hour.

Plant

A combination of hired plant and company plant will be used for the contract. The rates are shown in Table 5.2.

Materials

The quotations shown in Table 5.3 were received for the materials required for the construction of the works.

Table 5.1 :   Labour rates

Category	Rate per hour
	£
Labourer	5.10
Steel fixer	6.30
Carpenter	6.30
Plant operator	7.27
Ganger	6.69
Chargehand carpenter	6.90
Pipe layer	5.32

Table 5.2 :   Plant rates

Plant item	Rate per hour
	£
Caterpillar D8	39.11
22 RB Tracked Crane	16.83
Caterpillar 977 Dozer	26.94
16-tonne truck or lorry	25.00
Compressor 2 C.m/min	3.61
Tractor and trailer	10.25
JCB 3CX	19.90
TS24 Scraper	52.98
Grader	41.96
Caterpillar D6	32.22
Caterpillar D7	56.22
72 T Roller	5.40
Hymac excavator	21.40
941 excavator	22.10
Drott	14.37
Bomag roller	6.28

Table 5.3 :   Materials quotations

Materials	Quote
	£
Mild steel rebar BS 4449:	
10 mm diameter MS rebar	437.45 per tonne
12 mm " "	438.11 "
16 mm " "	405.11 "
20 mm " "	405.11 "
25 mm " "	406.27 "
32 mm " "	416.70 "
40 mm " "	425.74 "
Timber:	
Timber for falsework	253.40/m^3

18 mm ply	6.59/m²
Concrete:	
Lean mix	26.65/m³
Grade C15 OPC	30.20/m³
Grade C20 OPC	49.45/m³
Grade C25 OPC	51.63/m³
Grade C30 OPC	53.36/m³

EARTHWORKS

The earthworks within a civil engineering contract often represent a significant proportion of the total contract cost. The construction method and the outputs that can be achieved depend on:

* the type of material;
* the volumes involved;
* the source of the material; and
* the plant available.

The estimator is required to consider all these factors in calculating the cost of the earthworks.

Type of material

The type of material that must be moved will determine the equipment to be used, the rate of loading and the method of transportation. The estimator must carefully study the borehole information provided by the client's advisors and other available information such as geological maps. Wherever possible, samples of the material should be obtained. For the purposes of estimating, soil may be divided into the following categories:

* loose sand or loose soil;
* firm sand or firm soil;
* clay or heavy soil;
* compact soil;
* soft rock; and
* hard rock.

The characteristics of the material must be understood in order that a suitable allowance may be made for overbreak during excavation or the bulking of materials or any abnormal soil conditions such as running sand.

Earthwork volumes

The volumes used in earthworks calculations vary according to the material and its state.   The three volumes that can be calculated are:

* bank volume;
* bulked volume;   and
* compacted volume.

Bank volume (or in-place or solid volume) represents the soil in an undisturbed condition.   When soil is disturbed, i.e. on excavation, swelling or bulking takes place to give the *bulked volume*

bulked volume =   bank volume x swell factor.

When soil is compacted a different volume, the *compacted volume*, is obtained.   The compacted volume is calculated from the bank volume using a compaction factor:

compacted volume = bank volume x compaction factor.

This assumes that the material was compacted from a bulked condition. This is summarised for different types of soil in Table 5.4.

Table 5.4 :   Earthwork volumes

Materials	Bank volume	Bulked volume		Compacted volume	
Broken hard rock	1.0	1.5 -	2.0	1.3 -	1.4
Broken soft rock	1.0	1.5 -	2.0	1.3 -	1.4
Clay or heavy soil	1.0	1.3 -	1.4	0.8 -	1.0
Loose sand	1.0	1.0 -	1.1	0.9 -	1.0
Firm sand	1.0	1.1 -	1.3	0.9 -	1.0
Compact soil	1.0	1.1 -	1.3	0.8 -	0.9

The bank volume represents the 'paid' excavation volume as stated in the bill of quantities. The bulked volume represents the volumes to be transported and the compacted volume represents the fill in place. Invariably the bill of quantities assumes that the bank volume of material is the same as the compacted volume. If the estimator believes this not to be the case, then he must make adjustments to the figures used in his calculations.

Source of material

The material to be used as fill will be excavated and transported from a number of different sources including:

* the main carriageway of the road;
* the foundations to structures;
* accommodation works;
* drainage trenches; and
* imports from borrow pits or quarries.

Each of the sources will have different restrictions on the type of plant that can be used and this will vary the production rates that will be achieved.

Labour and plant

The estimator must examine the different combinations of labour and plant that are suitable for the work and estimate the cost of these combinations to produce the optimum configurations. The choice of plant may be restricted by that available from the company's own plant holding or from specialist plant hire companies if required by the specialist nature of the work involved. The plant hire companies in the area of the site must be contacted to determine both the availability and price of their plant. Where special items are required for large contracts it may be necessary to check throughout the country for their availability. Where such items are required an allowance must be included for their transportation to and from site.

EXCAVATION ESTIMATING EXAMPLES

A description of estimating examples is given together with basic estimating data for:

* manual excavation;

* excavation over a small site to reduced levels;
* large volumes of earthworks requiring cut and fill calculations; and
* excavation for structures.

The loading and carting away of materials are normally included as part of the examples.

Manual excavation

Manual excavation may be required to trim areas of formation or for trenches, small pits and isolated areas adjacent to structures.   The estimate will frequently be based on unit rate methods with a calculation of the number of hours per m³ (bank) required to excavate the material.   Typical values for various soil conditions are shown in Table 5.5.   An extra allowance of 1 hour/m³ should be made if the material is not to be loaded immediately into a barrow or skip and double handling is involved.

Example

The estimator is required to estimate the cost of 1.5 m³ of excavation relating to the connection of a section of pipework to the existing drainage. The work will require hand digging because of the existence of the drainage pipe and other services. The soil type is London clay. Double handling of the clay will be necessary before the soil is removed from the site by one of the dumpers provided for general site use.

Table 5.5 :   Excavation by hand in hours/m³ (bank)

Type of excavation	Loose sand or loose soil	Firm sand or firm soil	Clay or heavy soil	Compact soil	Soft rock
Reduce level	1.60	2.00	3.00	2.40	2.60
Trenches	1.86	2.33	3.49	2.80	3.03
Small pits	2.12	2.65	3.97	3.18	3.44
Small isolated holes	2.60	3.25	4.87	3.90	4.22

From Table 5.5:

Labourer hours required   = 3.97 hours

Double handling allowance = 1.00 hours

Total labourer time   = 3.97 + 1.00   = 4.97 hours

Labour rate = £5.10/hour

Therefore cost of excavation = 4.97 x £5.10   = £25.35 per m³.

(The cost of the dumper would be allowed for in the site overheads or on-costs.)

Excavation over a small site to reduced levels

Excavation over a small site to reduced levels is also normally calculated on a unit rate basis. The estimator will assume some form of front loading tracked excavator is being used and tabulate values for the labour and plant hours based upon varying average depths. It may be required to excavate over the site to reduced levels and push the material into spoil heaps or to excavate over the site to reduced levels and load the material into wagons.

Figs. 5.2 and 5.3 give typical outputs for a Drott excavator with a 0.6 m³ shovel. The labour hours are based on one banksman. Where it is considered necessary to include other labour an additional allowance should be made.

Depth of Material (in mm)	TYPE OF MATERIAL				
	Loose Sand or Loose Soil	Firm Sand or Firm Soil	Clay or Heavy Soil	Compact Soil	Soft Rock
50	0.0033	0.0044	0.0051	0.0067	0.0111
100	0.0067	0.0088	0.0102	0.0133	0.0221
150	0.0099	0.0133	0.0153	0.0200	0.0332
200	0.0133	0.0177	0.0204	0.0265	0.0442
300	0.0200	0.0265	0.0306	0.0398	0.0665

Fig. 5.2   Excavation oversite : plant hours/depth of material; material dozed into spoil heaps

Example

It is required to excavate 58 m² topsoil of average depth 300 mm over the area to be used for the construction of a culvert. The topsoil will be pushed into heaps and used to cover the embankments. It is decided to use a Drott with one labourer in attendance as a banksman.

From Fig. 5.2: Hours per square metre = 0.029 hour

$$\text{Cost of Drott} \qquad = \pounds 14.37/\text{hour}$$

$$\text{Cost of labourer} \qquad = \pounds 5.10/\text{hour}$$

Labourer rate for excavation = 0.029 x £ 5.10 = £ 0.15/m²

Plant rate for excavation   = 0.029 x £14.37 = £ 0.42/m²

Total rate for bill item   = £0.15 + £ 0.42 = £ 0.57/m²

Depth of Material (in mm)	TYPE OF MATERIAL				
	Loose Sand or Loose Soil	Firm Sand or Firm Soil	Clay or Heavy Soil	Compact Soil	Soft Rock
50	0.0025	0.0033	0.0037	0.0051	0.0074
100	0.0049	0.0067	0.0074	0.0101	0.0147
150	0.0074	0.0099	0.0111	0.0153	0.0221
200	0.0098	0.0132	0.0147	0.0204	0.0295
300	0.0147	0.0199	0.0221	0.0290	0.0442

TYPE OF MACHINE   :   DOZER

Fig. 5.3 Excavation oversite: plant hours/depth of material; material loaded into wagons

Large volumes of earthworks requiring cut and fill

When estimating for large volumes of earthworks a thorough understanding of the materials to be moved, the volumes involved, and the distances the material must be transported is needed. It is necessary to determine the overall nature of the project, i.e. whether the job is mostly cut and fill, mostly imported fill from borrow pits, or purchased fill from existing pits or quarries.

All available information on the material to be moved should be studied and wherever possible soil samples obtained. Possible fill materials must be tested to determine their suitability and the compactive effort required when the fill is used in embankments or as backfill to structures.

Material categories used in the calculations

The estimator must calculate from the quantities in the bill the total volume of material to be excavated and placed as fill. This is separated into the categories shown in Table 5.6.

Table 5.6:   Excavation and fill

Excavation -	Topsoil
-	Unsuitable material
-	Suitable material
-	Rock
Fill -	Suitable fill
-	Rock fill
-	Capping layer

The topsoil must be excavated, stored in a tip and then used as topsoil on the final embankments. Unsuitable material must be removed from site to an appropriate tip. Suitable material and rock that is excavated may be used to fill elsewhere in the works. Where the proposed formation is of insufficient strength the engineer may specify a capping layer of material.

Ideally the excavated quantity on the project should balance the required fill quantity. Where this is not the case, excess material must be brought on to site or disposed of to tip according to the discrepancy. Suitable sites for the tipping of materials or for use as borrow pits must be found and negotiations entered into to secure their use. All potential problems with the use of such sites must be investigated. Details of all statutory planning restrictions and any other relevant requirements should if possible be obtained at this stage. Where large volumes of imported fill material are needed for the project it is important that actual bulked and compacted volume figures for the material are available for the earthworks calculations. These should be obtained from similar contracts that have already been completed.

Cut and fill

The estimator must study the volumes of material and their location to determine the optimum movement of material on the contract. In practice it is necessary to perform the calculations for the total length of the road. For simplicity the following calculations reflect only a section of the works. It is necessary to tabulate for each area of cut and fill the volumes of material involved. An example is shown in Fig. 5.4.

This information may be shown diagrammatically to give a better appreciation of the possible movements of excavation and fill material. An example is given in Fig. 5.5. A horizontal line is drawn to scale to represent the length of the road. The volumes of cut and fill are then plotted with the excavation drawn above the line and the fill below. These volumes are colour coded and labelled with alphabetical characters for easy identification. The locations of structures and culverts are added to the diagram as these may affect the ability to move material during particular times on the contract. Suitable tips and borrow pits are located so that the distance to them from various chainages may be calculated.

Calculating the average lead distance

The average lead distance for each category of material is calculated from the volumes in each cut and fill location and the distance they are required to be transported. Having determined this average lead distance the estimator is able to select suitable plant configurations and calculate the probable cost per cubic metre for the excavation or placing and compaction of the material.

This is demonstrated in Tables 5.7 to 5.9 where the maximum overall average lead distance for the excavation of suitable material and its transportation to fill in the embankments and to tip is calculated.

CHAINAGE M	EXCAVATION				FILL			CHAINAGE MID POINT	DISTANCE
	TOPSOIL	UN-SUITABLE	SUITABLE	ROCK	SUITABLE	ROCK	CAPPING LAYER		
	m³	m³	m³	m³	m³	m³	m³	m	m
0000 – 0600	2600		A 47000					300	600
0600 – 1350	2200		B 40000					975	750
1350 – 2100	2500		C 18000					1725	750
2100 – 3200	4800		D 12000					2650	1100
3200 – 3950	2000					E 120000		3575	750
3950 – 4600	4400							4275	650
4600 – 5950	5600		F 41000					5275	1350
5950 – 6610	2080					G 20000		6280	660
TOTAL m³	26180		158000			140000			

Fig. 5.4 Cut and fill volumes for a section of the earthworks

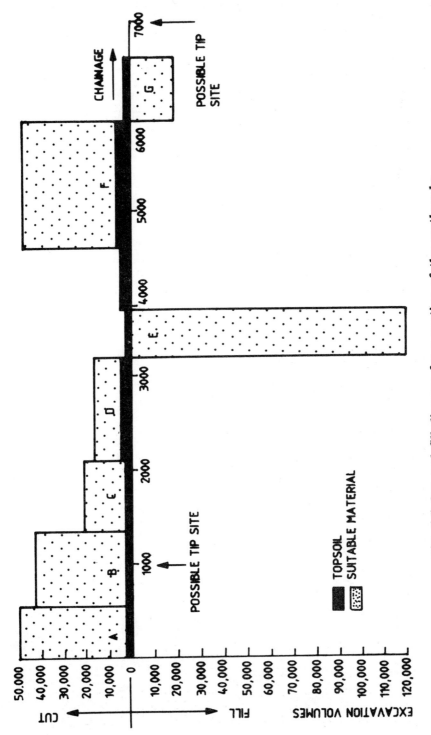

Fig. 5.5 Cut and fill diagram for a section of the earthworks

Table 5.7:   Excavation and transportation material to tip

From Figs 5.4 and 5.5 it is evident that

Total volume to be excavated	=	158000 m³
Total volume to fill	=	140000 m³
Material to tip	=	158000 - 140000 m³
	=	18000 m³

From Table 5.7 158000 m³ of suitable material will be excavated, and 18000 m³ of material must be taken to tip; 140000 m³ will be placed as fill in various areas. Two possible tip sites are available at chainages 1000 and 7000. The tip site giving the minimum overall average lead distance must be found.

Table 5.8:   Solution 1.   Use tip at chainage 1000

A	B	C	D
Haul	Vol. (m³)	Distance (m)	Vol. x dist. (B x C)
A to tip	18000	700	12600000
A to E	29000	3275	94975000
B to E	40000	2600	104000000
C to E	18000	1850	33300000
D to E	12000	925	11100000
F to E	21000	1700	35700000
F to G	20000	1005	20100000
Total	158000		311775000

Average lead = 311775000/158000 = 1973 metres

Table 5.9:   Solution 2.   Use tip at chainage 7000

A Haul	B Vol. (m³)	C Distance (m)	D Vol.x dist. (B x C)
A to E	47000	3275	153925000
B to E	40000	2600	104000000
C to E	18000	1850	33300000
D to E	12000	925	11100000
F to E	3000	1700	5100000
F to G	20000	1005	20100000
F to tip	18000	1725	31050000
Total	158000		358575000

Average lead = 358575000/158000 = 2269 m

Clearly Solution 1 is the better alternative as it gives less of an average lead.   For the earthworks on a motorway contract the minimum movement of material may only be found after analysing several alternative solutions. Having established the minimum average lead times, suitable plant configurations are selected and their costs calculated.

Here on

Selection of plant configurations and the calculation of their cost

The examples shown in Tables 5.10, 5.11 and 5.12 give calculations for :

* the excavation and transportation of topsoil;
* the excavation and transportation of suitable material;   and
* the compaction of fill material in embankments.

In each of these situations only a single plant configuration is proposed. If alternative solutions were available then they would be costed and compared to find the cheapest configuration.   Alternative solutions would also normally be examined where the average lead distance was close to the maximum acceptance for the plant configuration chosen.

Table 5.10:   Plant for excavation and transportation of topsoil

---

Total volume	=	26180 m³
Average lead distance	=	600 m

(Several suitable tip sites are available along the line of the road.)

Plant selected : TS24 scrapers at average speed 21.25 km/hour

Cycle time	: Manoeuvre	= 1.00 min
	: Load	= 1.50 min
	: Haul 0.6 x 60/20	= 1.80 min
	: Manoeuvre and unload	= 2.50 min
	: Return 0.60 x 60/22.5	= 1.60 min
	Total cycle time	= 8.40 min

It is assumed that the speed of the scraper is 20 km per hour loaded and 22.5 km per hour unloaded.

Total cycle time assumed	=10 min per round trip.
Carrying capacity of TS24	=18 m³

Bulking is assumed to be allowed for by the heaping of material. Underloading will be prevented by push loading with a dozer.

Manoeuvre and load time   =1.00 + 1.50 = 2.5 min.

Number of TS24 scrapers required to keep D8 pusher constantly at work is then = cycle time/(manoeuvre + load time).

No. required = 10/2.5 = 4 scrapers. This is shown graphically in Fig. 5.6.

Cost of plant : TS24 scraper £52.98 per hour; D8 pusher £64.64 per hour

Total plant group cost   =   (4 x £52.98) + £64.64

                         =   £276.56 per hour.

Cost of plant group per 50 hour week = £276.56 x 50 = £13828.00

Total volume hauled per week : number of scrapers x volume carried x trips per hour x hours worked = 4 x 18 x 6 x 50 = 21600 m³.

Plant cost per cubic metre   = 13828.00/21600 = £0.64/m³

Cost of labour   = ganger @ £6.69 per hour

Cost of ganger per 50 hour week = 50 x £6.69 = £334.50.

Volume hauled per week = 21600 m³

Therefore labour cost   = £334.50/21600 = £0.015/m³

Total rate - excavation and transportation of topsoil

= labour rate + plant rate = 0.015 + 0.64 = £0.655/m³

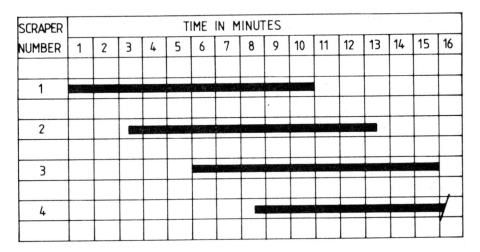

Fig. 5.6 : TS24 requirements for the excavation and transportation of topsoil

Table 5.11: Excavation and transportation of suitable fill to embankments

Total volume	=	158000 m³
Average lead distance	=	1973 m (from Table 5.8)

Plant selected : TS24 scrapers which will travel at an average speed of 21.25 km/hour carrying capacity 18m³

Cycle time	: Manoeuvre	=	1.50 min
	: Load	=	1.50 min
	: Haul 1.973 x 60/20	=	5.92 min
	: Road crossing delays	=	1.50 min
	: Manoeuvre and unload	=	2.50 min
	: Return 1.97 x 60/22.5	=	5.25 min
	Total cycle time	=	18.17 min

Assume total cycle time = 20 min, i.e. 3 trips per hour for each TS24. Bulking of material allowed for by the reduced volume in each load plus heaping of material. Underloading will be prevented by push loading with a dozer.

Manoeuvre and load time   =   1.5 + 1.5 = 3.0 min.

Number of TS24 scrapers required to keep D8 load constantly at work

= cycle time/(manoeuvre + load time) = 20/3.0 = 6.67.

Allow for 6 scrapers with 6 min per hour for the D8 loader to manoeuvre. A grader will be required to clear the haul road.

Cost of plant : TS24 scraper   £52.98 per hour

              : D8 loader   £64.64 per hour

              : grader   £41.96 per hour

Total plant group cost = (6 x £52.98) + £64.64 + £41.96

                        = £424.48 per hour.

Cost of plant group per 50-hour week = £424.48 x 50 = £21224.00

Total volume hauled per week = number of scrapers x volume carried

                               x trips per hour x hours worked

                          = 6 x 18 x 3 x 50   =   16200 m^3

Plant cost per cubic metre   = £21224/16200   = £1.31/m^3

Cost of labour : ganger   @ £6.69 per hour

              : labourer   @ £5.10 per hour

Cost of labour gang per 50-hour week = 50 x £11.79   = £589.50

Labour cost per m^3 hauled = £589.50/16200 = £0.036/m^3

Suitable rate for excavation and transportation of suitable fill to the embankments = labour rate + plant rate = 0.036 + 1.31 = £1.35/m^3.

Table 5.12:   Compaction of fill material to embankments

Plant will be required to spread and compact fill material deposited by TS24 scrapers.

Available plant: Caterpillar D7 or Komatsu D65A.

The Caterpillar D7 is selected on account of the cheaper hire rate.

For compaction assume 72 T roller towed by a D6 dozer at 8 km per hour. Width of the roller is 1.7 m.

Area rolled per hour =   8 x 1000 x 1.7   =   13600 m^2

Area rolled per week =   50 x 13600   = 680000 m^2

Maximum rate of deposited fill material (from Table 5.11) = 16200 m^3

Assume a depth of 0.4 m.

This represents an area of 16200/0.4 = 40500 m^2 of material deposited per week.

Therefore allow for one 72 T roller towed by a D6 dozer.

Cost of plant: D7 to spread material   @   £ 56.22 per hour

D6 to compact material @   £ 32.22 per hour

72 T roller   @   £   5.40 per hour

Total cost of plant group per hour   =   £ 93.84 per hour

Cost of plant group per 50-hour week   =   £93.84 x 50 = £4692.00

Cost of plant group per cubic metre compacted

$$= \quad £4692.00/16200 \text{ m}^3$$

$$= \quad £0.29/\text{m}^3$$

Labour cost: 1 ganger required to supervise compaction @ £6.69/per hour

Cost per 50-hour week   = 50 x £6.69   =   £334.50

Cost per cubic metre of material   = £334.50/16200 m³ = £0.02/m³

Total rate for the compaction and fill of material to embankment

= labour rate + plant rate =   0.02 + 0.29   =   £0.31/m³.

Excavation for structures

Each different structure on a project should be treated as a special case and appropriate labour and plant allowances calculated from consideration of the volume of material to be excavated and the working space available. Some standard outputs may be assumed for normal trunk road or motorway bridges, e.g. a multispan in-situ slab or precast beam deck structure supported by columns or piers with abutments and wingwalls constructed on normal spread footings. Excavation for these types of structure is normally undertaken with a tracked backacter type excavator.

Fig. 5.7 gives bulked production volumes for a backacter type excavator working in three different types of dig: easy, medium and hard. The figures are given for different capacities of bucket. Adjustment factors are also included in order that the production figures may be used for other types of excavating machinery.

The output figures relate to 'paid' excavation. It is important to consider the overbreak volume of material as this will have to be laid aside for later backfilling. The actual overbreak allowance will depend upon the soil type, the depth of the dig and the working space required.

The example in Table 5.13 relates to the excavation of suitable material to the structural foundation of a typical motorway overbridge.

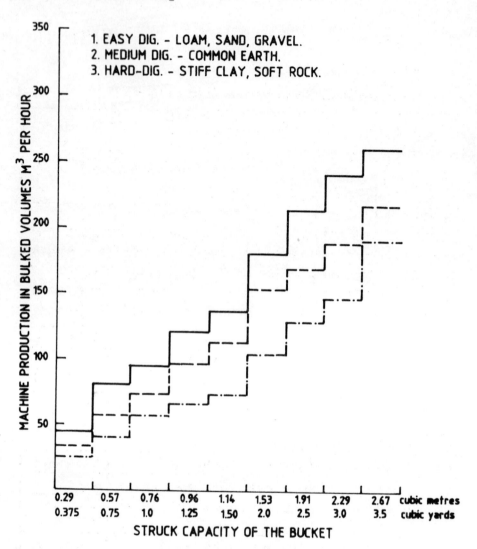

Fig. 5.7   Bulked production volumes for a backacter excavator
from reference (3)

DRAIN GROUPS

Surface water drains

Pipe and bed combinations permitted in the contract.

Pipe Material		CLAY			CONCRETE			ASBESTOS CEMENT				UPVC			Ductile Iron
Strength Class Dia.	Group No.	Standard	Extra	Super	Standard	Medium	Heavy	Light	Medium	Heavy		B	C	D	
150	4	Z	Z	Z	Z					Z					
150	8	ABN	ABN	ABN	ABN					ABN		E	E	E	Q
225	9	AB	ABN		ABN					ABN		E	E	E	Q
225	10	Z	Z		Z					Z					
300	7	AB	ABN		AB		AB		ABN	ABN		E	E	E	Q
300	8	AB	ABN		AB		ABN		ABN	ABN		E	E	E	Q
300	10	A	AB		AB		AB		AB	ABN		E	E	E	Q
300	11	Z	Z		Z		Z		Z	Z					
375	4	Z	Z		Z	Z			Z	Z					
375	7	AB	ABN		AB	ABN	ABN		ABN	ABN		E	E	E	
375	8	AB	ABN		AB	AB	ABN		ABN	ABN		E	E	E	
375	13	AB	AB		A	AB	AB		AB	ABN		E	E	E	
460	7	AB	ABN		A	ABN	ABN	AB	ABN	ABN		E	E	E	E
450	8	AB	ABN		A	AB	ABN	AB	ABN	ABN		E	E	E	E
450	10	AB	AB		A	AB	AB	AB	AB	ABN		E	E	E	E
450	11	Z	Z		Z	Z	Z	Z	Z	Z					
600	7					ABN	ABN	ABN	ABN	ABN		E	E	E	
600	10					AB	ABN	ABN	AB	ABN		E	E	E	Q
600	13				Z	Z	Z		Z	Z					

Fig. 5.8 Specification of pipe and bedding combinations

Table 5.13:   Excavation of suitable material in structural foundations

---

Total volume to be excavated   =   36.00 m^3

Depth of excavation varies between 0 and 3.0 m.

Use a plant and labour gang consisting of:

      1 Hymac excavator @ £21.40 per hour.

      2 labourers   @ £5.10 per hour.

One labourer will act as banksman to the machine and the other will trim the formation.

Total gang cost per hour   =   £21.40 + (2 x £5.10)   =   £31.60.

Assume an output rate of 11.5 m^3 per hour.

Cost of excavation   =   £31.60/11.5   =   £2.75/m^3

Allowance for overbreak   =   100%

Therefore corrected excavation rate   =   £5.50/m^3

The excavated material must be hauled 160 m to a suitable fill location.

Use 10 tonne tippers transporting 4.0 m^3 per cycle.

Assume a haulage speed of 6 km/hour

Loading time	=	(4/11.5) x 60	=	21.0 min
Haul time	=	(160/6000) x 60	=	1.6 min
Tipping time			=	1.0 min
Return time			=	1.6 min
Total cycle time	=	21 + 1.6 + 1.0 + 1.6	=	25.2 min

Assume 3 wagons to allow for breakdowns, etc.

Cost of wagon   =   £19.47

Cost of transport =   3 x (19.47/11.5)   =   £5.08/m^3

Total cost of excavation and transportation   =   £5.50 + £5.08

                                          =   £10.58/m^3.

DRAINAGE

The following types of estimate for drainage contracts commonly occur in civil engineering contracts:

    *   storm and foulwater drains;

    *   french drains;

    *   drainage to the toe of embankments and the top of cuttings;

    *   connections to both new and existing drainage;

* manholes, catchpits and gullies;   and
* piped culverts.

STORM AND FOUL WATER DRAINAGE

It is necessary to calculate the labour,   plant and material required for each different length of drainage within the contract.

From the drawings, specifications and other contract documents the estimator will determine:

* the pipe and bedding combinations;
* the nature of the ground in which the trenches will have to be dug; and
* any major obstructions to the drainage water to be undertaken.

The estimator then has to decide the labour and plant groups that will be required and their probable output rates.

The specification gives the pipe and bedding combinations permitted within the contract and a typical example is given in Fig. 5.8.   For the diameters of various types of pipe a reference to the type of surround detail is given which must be read with the appropriate drawing.

From this sheet the estimator may select the type of construction that will be cheapest based upon the quotations received for pipes and bedding materials.   The determining factor may vary with the exact detail but the cost of the pipe is usually the most important factor.   For simplicity of construction, ease in the purchasing of materials and speed in performing the estimating calculations, the minimum number of pipe and bedding combinations are usually chosen.

Also contained in the specification will be a full schedule of the sections of drainage abstracted from the drawings and a schedule of manholes and catchpits. This enables the estimator to augment the details of drainage given in the bill and obtain a complete picture of the requirements of the contract. Examples of these are given in Figs 5.9 and 5.10.   Usually the drainage schedule lists each drainage 'run' and gives details of:

* the type of drain;
* the location;
* the length in metres;
* the diameter of the pipe;
* the design group;

ITEM No.	LOCATION		LENGTH METRES	TYPE	DIA (mm)	GROUP	DEPTH(mm)		INVERT LEVELS		GULLY CONN.
	From	To					Av	Max	From	To	
	2078	CP.24	72	FD.	150	1	1425	1500	32.91	31.43	2
	CP.24	CP.25	25	FD.	150	1	1490	1500	31.43	30.92	
	2078	2109	31	FD.	150	1	1425	1500	33.08	32.32	
	2108	CP.26	67	FD.	150	1	1650	1840	32.32	30.62	
	2078	CP.27	97	FD.	150	1	1430	1500	32.91	30.93	2
	2178	2197	19	FD.	150	1	940	1000	31.05	30.65	
	2197	CP.28	43	FD.	150	1	1650	1830	30.65	29.75	
	2300	CP.25A	114	FD.	150	1	1035	1200	25.13	24.00	
	2347	2290	57	FD.	150	1	1330	1500	30.10	29.91	
	2290	CP.28	50	FD.	150	1	1685	1830	29.91	29.75	
	2447	CP.29	97	FD.	150	1	1160	1180	31.00	30.19	
	CP.30A	CP.30B	40	FD.	225	1	1200	1300	28.50	27.90	
	CP.30B	CP.30C	55	FD.	225	1	1200	1300	27.90	27.40	
	CP.35B	CP.32A	30	FD.	225	1	1100	1200	35.00	27.80	
	CP.32A	CP.32B	75	FD.	225	1	1250	1300	27.80	27.20	
	CP.32B	CP.32C	75	FD.	225	1	900	1300	27.20	27.00	
	CP.24A	CP.24	16	SWD	150	4	930	1280	36.80	31.43	
	CP.25	CP.26	15	SWD	150	4	1055	1100	30.92	30.62	
	CP.26	CP.27	15	SWD	150	4	1050	1100	30.62	30.48	
	CP.25A	MH.19A	63	SWD	150	4	1300	1350	24.00	23.50	
	MH.18	CP.28	15	SWD	150	4	625	750	30.03	29.63	
	CP.28	MH.19	13	SWD	150	4	1300	1490	29.68	29.57	
	MH.20	CP.29	15	SWD	150	4	1000	1180	30.29	30.19	
	CP.29	MH.21	14	SWD	150	4	1190	1200	30.19	30.01	
	CP.30	CP.31	9.5	SWD	150	4	1830	1970	30.56	30.46	
	CP.31A	CP.32	10.5	SWD	150	4	2060	2100	30.46	30.36	
	CP.23	CP.27	100	SWD	300	7	2030	2100	32.31	30.33	
	CP.27	MH.19	65	SWD	300	7	1510	1600	30.33	29.50	1

Fig. 5.9   Schedule of drains

* average and maximum depths;
* invert levels;
* gully connections (where appropriate);   and
* the relevant drawing number.

The schedule of catchpits and manholes usually gives information on:

* the reference number of the manhole;
* the depth to invert;
* the relevant drawing number;   and
* the number and type of gully connections.

Where these details are not provided they must be abstracted by the estimator from the drawings. The drainage details taken from the specification and the bill of quantities are entered on to a separate summary sheet, which is used to estimate the cost of the drainage for a contract. An example is shown in Fig. 5.11.

Examples of typical calculations for storm and foulwater drains

Materials requirements

From the drainage details the estimator has determined that two types of drainage bedding will satisfy all the requirements for the different diameter pipe runs on the contract. These are shown in Figs 5.12 and 5.13.

For each pipe diameter the volume of bedding material per metre run of trench is calculated. This is dependent on the width of the base of the trench. Although the specification determines the minimum and maximum widths allowed, for smaller sized pipes (less than 300 mm internal diameter) a minimum width equivalent to the bucket of the excavating machine is normally taken. For larger pipes a width of trench equal to the external diameter of the pipe + 450 mm is typically assumed.

300 mm pipe

BEDDING DETAIL TYPE Z

Width of trench	=	750 mm
External diameter of pipe	=	0.334 m
Depth of bedding	=	100 mm (minimum)
Cover to pipe	=	150 mm

SCHEDULE OF CATCHPITS & manholes

DESIGN GROUP	DEPTH CLASS	REF NO.	DEPTH (metres)	DRG NO A1131-	GULLY CONN No.	GULLY CONN TYPE	No. OF CONNECTIONS 150mm	225mm	300mm	375mm	450mm	600mm	1050mm	NOTES
CATCHPITS														
	2m or less	CP.43A	1.850	DR8			2							
		CP.43B	1.475	DR8			3							
		CP.43C	1.500	DR8			2							
		CP.49	1.395	DR8			2							
		CP.49A	1.500	DR8			2							
		CP.52	1.705	DR8	1		2							
	2m–3m	CP.43	2.195	DR8			3							
		CP.47	2.760	DR8			2	1						
		CP.50	2.640	DR8				2						
		CP.51	2.020	DR8			2	1						
	3m–4m	CP.48	3.185	DR8	1		1		2					
MANHOLES														
	2m or less	MH.23	1.565	DR8	1		1	1						
		MH.24	1.450	DR8	1		2	1						
		MH.87	1.985	DR8	1			3						
	2m–3m	MH.22	2.380	DR8				2	1					

Fig. 5.10   Schedule of catchpits and manholes

CONTRACT: LUTON EAST CIRCULAR — ESTIMATING SHEET No/

MAIN CARRIAGEWAY DRAINAGE AND SERVICE DUCTS

BILL REF	PIPE DIA.	MH-MH	LOCATION	MISC	depth (m)	length (m)	m³/m	total m³	EXCAVATION rate m³	EXCAVATION rate m	BED OR BED & SURROUND m³/m	rate m³	rate m	PIPE RATES lay m	pipe m	rate for reinstate m	NETT TOTAL RATE	L	P	M
501	150	G.P.1		G.b.h.	1·50	148	1·13		2·08	2·35	·122	8·92	1·09		2·43		8·13	2·35	2·26	3·52
502	150	G.P.3		G.b.h.	1·50	15	1·13		2·08	2·35	·122	8·92	1·09		2·43		8·13	2·35	2·26	3·52
503	150	G.P.3		G.b.h.	1·64	20	1·23		2·08	2·56	·122	8·92	1·09		2·43		8·54	2·56	2·46	3·52
505	150	G.P.4		G.b.h.	1·50	188	1·13		2·08	2·35	·122	8·92	1·09		2·43		8·13	2·35	2·26	3·52
506	150	G.P.4		G.b.h.	1·68	3	1·26		2·08	2·62	·122	8·92	1·09		2·43		8·30	2·26	2·52	3·52
508	150	E.O. FOR Z				70	—								2·43		2·43	2·43	2·43	2·43
509	150	G.P.12		G.b.h.	2·40	26	1·80		2·08	3·74	·122	8·92	1·09		2·43		10·86	3·74	3·60	3·52
511	225	G.P.1+Z		G.b.h.	1·50	37	1·13		2·08	2·35	·595	20·67	12·30		4·83		21·74	2·35	2·26	17·13
512	225	G.P.4.		G.b.h.	1·50	754	1·13		2·08	2·35	·140	8·92	1·25		4·83		10·69	2·35	2·26	6·08
513	225	G.P.4.		G.b.h.	1·74	35	1·31		2·08	2·72	·140	8·92	1·25		4·83		11·42	2·72	2·62	6·08
515	225	G.P.4+Z		CONC. SUR	1·50	44	1·13		2·08	2·35	·595	20·67	12·30		4·83		21·74	2·35	2·26	17·13
516	225	G.P.5+Z		CONC. SUR	1·50	14	1·13		2·08	2·35	·595	20·67	12·30		4·83		21·74	2·35	2·26	17·13
517	225	G.P.10+Z		CONC. SUR	1·50	14	1·13		2·08	2·35	·595	20·67	12·30		4·83		21·74	2·35	2·26	17·13
518	300	G.P.3		Gbh	1·50	209	1·58		2·08	3·29	·230m³	8·92	2·05		7·94		16·44	3·29	3·16	9·99
519	300	G.P.3	1·05	Gbh	1·50	7	1·58		2·08	3·29	·230	8·92	2·05		7·94		16·44	3·29	3·16	9·99
520	300	G.P.3.	SLOPES	Gbh	2·04	24	2·14		2·08	4·45	·230	8·92	2·05		7·94		18·72	4·45	4·28	9·99
522	300	G.P.3+Z		CONC. SUR.	1·50	27	1·58		2·08	3·29	·595	20·67	12·30		7·94		26·69	3·29	3·16	20·24
523	300	G.P.3+Z		CONC. SUR	1·50	25	1·58		2·08	3·29	·595	20·67	12·30		7·94		26·69	3·29	3·16	20·24
524	300	G.P.3+Z		CONC. SUR	1·73	6	1·82		2·08	3·79	·595	20·67	12·30		7·94		27·67	3·74	3·64	20·24
526	300	G.P.3+Z		CONC. SUR	1·82	8	1·91		2·08	3·97	·595	20·67	12·30		7·94		28·03	3·97	3·82	20·24
528	300	G.P.4		Gbh	1·50	131	1·13		2·08	2·35	·230	8·92	2·05		7·94		14·60	2·35	2·26	9·99
529	300	G.P.4		G.b.h.	1·87	68	1·96		2·08	4·08	·230	8·92	2·05		7·94		17·99	4·08	3·92	9·99
531	525	G.P.11	1·28	G.b.h.(h)	1·50	11	1·13		3·75	4·24	·444	8·92	3·96		14·20		25·63	4·24	3·23	18·16
532	525	G.P.11	1·28	G.b.h.(h)	3·61	51	5·42		3·75	20·33	·444	8·92	3·96		14·20		53·79	20·33	1550	18·16
SE2																				

Fig. 5.11   Abstract and calculation sheet for drainage

Volume of concrete per metre run = overall depth x width

$\qquad$ - volume of pipe

$= (0.150 + 0.334 + 0.1) \times 0.75 - 3.142 \times (0.334 \times 0.334)/4$

$= 0.584 \times 0.75 - 0.088 = 0.35$ m³/metre run.

BEDDING DETAIL TYPE B

The volume of bedding material is assumed to be 0.5 x volume for concrete surround $= 0.175$ m³/metre run.

Selected backfill $\qquad$ = overall depth x width - volume of pipe

$\qquad$ - volume of bedding

$= (0.1 + 0.334 + 0.3) \times 0.750 - 0.088 - 0.175$

$= 0.550 - 0.088 - 0.175 = 0.287$ m³/metre run.

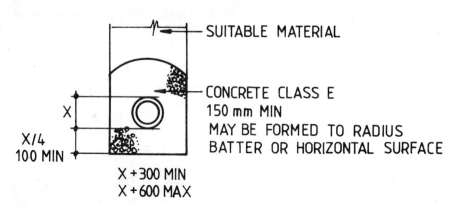

Fig.   5.12   Bedding detail type Z

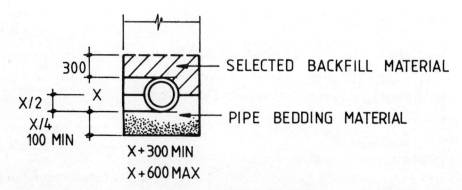

Fig.   5.13   Bedding detail type B

These calculations are then repeated for each different pipe diameter included in the contract.

Estimators usually have calculation tables giving the material requirements for different pipe diameters and different typical surround details. An example is given in Fig. 5.14.

The price of materials

Bedding material

The quoted granular bedding material is £9.20/tonne.

Conversion at 1.60 tonne per m³ gives a cost of £14.70/m³.

A wastage factor of 10% is assumed.

The concrete will be provided by 'Ready Mix' and site delivered.

A price of £46.83/m³ is quoted. The top surface of the concrete in the trench will be left horizontal.

A wastage of 5% will be assumed.

Pipes

The following prices for pipes are obtained from a quotation from a suitable supplier:

Concrete class		L 150	mm diameter	£ 8.60/m
"	"	L 225	mm "	£12.45/m
"	"	L 300	mm "	£16.47/m
"	"	M 375	mm "	£20.56/m
"	"	M 450	mm "	£25.00/m
"	"	M 600	mm "	£35.96/m

A wastage allowance of 3% is used to allow for breakages when off-loading the pipes and when laying the pipes in the trenches. It is assumed that there will be no double handling of materials so that transportation costs over and above those normally allowed for in the site on-costs need not be included.

From these costs and the quantities calculated for the different diameters of pipe with each type of surround detail the cost of material for a metre run of pipe may be calculated.

CUBIC METRES OF CONCRETE PER METRE OF PIPE FOR VARIOUS CROSS SECTIONS

NOTE. Concrete taken in all cases as being 150mm wider than outside diameter of pipes on each side

CROSS SECTION OF CONCRETE	INTERNAL DIAMETER OF PIPE IN mm AND VOLUME OF CONCRETE IN CUBIC METRES/m															
	150	225	300	375	450	525	600	675	750	825	900	975	1050	1125	1200	1275
(diagram)	0·065	0·085	0·108	0·132	0·166	0·211	0·257	0·311	0·380	0·456	0·539	0·630	0·733	0·839	0·951	1·063
(diagram)	0·072	0·099	0·127	0·158	0·198	0·252	0·307	0·368	0·448	0·538	0·635	0·742	0·864	0·989	1·121	1·251
(diagram)	0·128	0·170	0·214	0·261	0·319	0·392	0·468	0·549	0·666	0·800	0·943	1·102	1·283	1·467	1·666	1·855
(diagram)	0·130	0·171	0·216	0·264	0·323	0·397	0·473	0·555	0·674	0·809	0·954	1·115	1·299	1·485	1·683	1·876
(diagram)	0·047	0·062	0·079	0·097	0·125	0·164	0·206	0·252	0·308	0·371	0·438	0·512	0·596	0·682	0·774	0·865
(diagram)	0·040	0·048	0·059	0·065	0·082	0·108	0·136	0·167	0·204	0·246	0·291	0·340	0·396	0·453	0·514	0·576

Fig. 5.14 Pipe bedding and surround material requirements from reference (78)

Example estimate for materials

The following example shows the calculation for the material cost of a metre run of 300 mm diameter concrete pipe class L with surround detail type Z (see Fig. 5.12).   The cost of the material is calculated in Table 5.14.

Table 5.14:   Calculation of the material cost of a metre run of 300 mm diameter concrete pipe class L

Cost of pipe	=	£16.47 per metre run
Wastage factor	=	3%
Cost of concrete surround	=	£46.83/m³
Volume required per metre run	=	0.35 m³/m
Wastage factor	=	5%
Cost of pipe	=	1.03 x 16.47
	=	£16.96 per metre run
Cost of concrete surround	=	1.05 x 0.35 x 46.83
	=	£17.21 per metre run
Total material cost	=	£16.47 + £17.21
	=	£33.68 per metre run

Labour and plant requirements

Different labour and plant configurations are necessary depending on the type of ground, diameter of pipe and the depth of trench required.   Typical configurations with weekly costs are given in Tables 5.15 to 5.17.

The operations involved within a pipework item include:

- the provision of pipes and surround materials;
- the excavation of the trench;
- the laying and testing of pipework;   and
- the backfill and compaction of the trench.

Of these operations the excavation of the trench is normally the item of work that determines the production rate of the gang.   The cost of the labour and plant per metre run of trench may be calculated as follows:

(i)   list the bill items of each diameter pipe and their relevant quantities;

(ii)   calculate the average volume of excavation for a linear metre run of trench;

(iii) select a suitable labour and plant gang;

(iv)   estimate the average production rate per week; and

(v)   calculate the cost of labour and plant for each bill item based upon the volume per metre run of trench.

A typical calculation for the average volume of excavation per metre run of trench is shown in Table 5.18.

The average depth of each trench run may be determined from the contract drawings. The width of a drainage trench will vary with the following factors:

-    pipe diameter and surround detail;
-    the width of excavation bucket;
-    the depth of the trench;
-    the soil type and overbreak; and
-    the working space required at the base of the trench.

Average trench widths for excavation are given in Table 5.19.

Table 5.15: Labour and plant requirements - gang type A

This gang is an example of that typically used for pipe diameters up to and including 300 mm.

1 JCB 3CX excavator	@ £	19.90 per hour
Small plant (pump, compactor, etc.)	@ £	2.00 " "
1 Pipelayer	@ £	5.32 " "
2 Labourers	@ £	5.10 " "

Gang cost per hour -

= £19.90 + £2.00 + £5.32 + (2 x £5.10) =   £37.42

Assuming a 50-hour week, gang cost per week

= 50 x £37.42 = £1871.00

Table 5.16: Labour and plant requirements   -   gang type B

This gang is an example of that typically used for pipe diameters between 375 mm and 975 mm.

1 Hymac excavator	@ £ 21.40 per hour
1 Cat. 941 excavator (50% usage)	@ £ 22.10 " "
Small plant (pump, compactor, etc.)	@ £ 2.00 " "
1 Pipelayer	@ £ 5.32 " "
3 Labourers	@ £ 5.10 " "

Gang cost per hour

$= £21.40 + (0.50 \times £22.10) + £2.00 + £5.32 + (3 \times £5.10) = £55.07$

Assuming a 50-hour week, then gang cost per week

$= 50 \times £55.07 = £2753.50$

Table 5.17: Labour and plant requirements   -   gang type C

This gang is an example of that normally used for pipe diameters above 975 mm.

1 Hymac excavator	@ £ 21.40 per hour
1 Cat. 941 excavator	@ £ 22.10 " "
Small plant (pump, Bomag roller, etc.)	@ £ 4.00 " "
1 Pipelayer	@ £ 5.32 " "
1 Ganger	@ £ 6.69 " "
3 Labourers	@ £ 5.10 " "

Gang cost per hour

$= £21.40 + £22.10 + £4.00 + £5.32 + £6.69 + (3 \times £5.10) = £74.81$

Assuming a 50-hour week, then the gang cost per week

$= 50 \times £74.81 = £3740.50.$

Table 5.18:   Calculation of the average volume of excavation -
150 mm dia. pipe

Bill item	Linear metres	Width	Depth	Volume (m³)	Vol/metre run (m³)
5.8	631	0.65	1.50	615	0.97
5.9	5	0.85	1.60	7	1.36
5.11	274	0.85	2.00	466	1.70
5.13	15	0.65	1.50	15	0.97

Total volume of excavation   = 615 + 7 + 466 + 15 = 1103 m³
Total length of trench   = 631 + 5 + 274 + 15 = 925 lin m
Average volume per metre run = 1103/925 = 1.19 m³

Table 5.19:   Average trench widths

Depth of trench (m)		Average trench width (m)		
Up to	1.50	Pipe dia.	+	0.50
1.50	to 2.00	"	+	0.70
2.00	to 3.00	"	+	1.00
3.00	to 4.00	"	+	1.20
4.00	to 5.00	"	+	1.50
5.00	to 6.00	"	+	2.00
6.00	to 7.00	"	+	2.50
7.00	to 8.00	"	+	3.00

In this example the estimator decides to use gang type A for the work. The average production per week is dependent on the location of the work. (There will obviously be a great difference in production of the same gang in an urban environment compared with a 'green field' site of a new motorway.) In this situation the estimator decides the gang will lay on average 1 pipe per hour, say 50 linear metres per week. The labour and plant cost per metre may then be calculated as follows:

Length of pipe run : 631 m

Volume of excavation per metre run of pipe : 0.97 m³

Average volume of excavation per linear metre of pipe : 1.19 m³/m

Total labour and plant cost per week : £1871.00

Average output : 50 linear metres

Average cost per metre : (£1871.00/50) = £37.42

Labour and plant cost for run : (0.97/1.19) x £37.42   =   £30.50/m.

This calculation is then repeated for each item of the same diameter.

The material excavated from the trench may be used for backfill providing that it meets the necessary specification. This will require a suitable area to stockpile the material during drainage construction. Some allowance must be made for the carting away of unsuitable material. Where imported backfill material is necessary this cost must be calculated separately and added to the rate.

Sheeting and strutting to trenches

In trenches up to 1.5 m deep, sheeting and strutting is not normally required. For trenches of depth greater than 1.5 m the allowance may be made depending on the overall depth of the trench and the soil condition.

A common form of sheeting and strutting is shown in Fig. 5.15. Timber struts and timber or steel walers are used to support steel trench sheeting. Depending on the ground condition and the depths of trench the sheeting may be:

- close sheeted (the trench sheets interlocked or 'buttoned up');
- medium sheeted (alternate trench sheets omitted, leaving a gap of approximately 0.33 m between each);   and
- open sheeted (two trench sheets omitted leaving a gap of 0.66 m).

Unless specialist trench support materials are required the materials are usually assumed to be available on site as general auxiliary plant items, the cost of which is covered in the preliminary items. Consequently no allowance is included in the drainage items. The additional labour requirement may be calculated on a unit rate basis.

Typical requirements for various depths of trenches are shown in Table 5.20.

FRENCH DRAINS

French drains are designed with porous or perforated pipes. The trench is normally backfilled with filter material specified as Type A or Type B under Table 5.7 of the Specification for Highway Works.   Fig. 5.16 shows a typical detail for French drain construction.

The cost of the labour and plant to construct the drain may be calculated in the manner as shown for storm and foulwater drains. Because of the cost of the Type A or B filter material the material cost must be calculated separately for each drainage run.   An example is shown in Table 5.21.

Table 5.20:   Labour allowance for planking and strutting to drain trenches

Depth (m)	Total area (m²)	Labour hours/m run
1.75	3.5	0.20
2.00	4.0	0.30
2.25	4.5	0.45
2.50	5.0	0.70
2.75	5.5	0.95
3.00	6.0	1.25
3.25	6.5	1.50
3.50	7.0	1.80
3.75	7.5	2.15
4.00	8.0	2.45
4.25	8.5	2.75
4.50	9.0	3.15
4.75	9.5	3.50
5.00	10.0	3.85
5.25	10.5	4.25
5.50	11.0	4.65
5.75	11.5	5.05
6.00	12.0	5.50

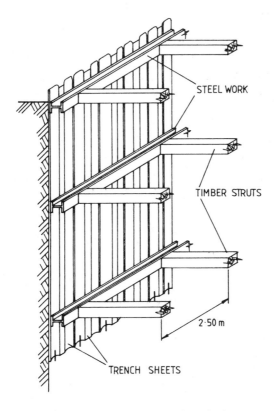

STEEL WORK

TIMBER STRUTS

2·50 m

TRENCH SHEETS

Fig. 5.15   Sheeting and strutting for drainage trenches

Table 5.21:   The calculation of material cost for a French drain

Length of pipe run   =   276 m

Average depth to invert   =   1.5 m

Pipe diameter   =   150 mm

Assumed trench width   =   650 mm

For consideration of the surround detail see Fig. 5.16.

Volume of granular bedding material Type A

= 0.075 x 0.650   =   0.049 m³ per metre run

Wastage allowance   =   5%

Cost of granular bedding material=   £14.70/m³

Cost of granular bedding material per metre run

= 1.05 x 0.049 x 14.70   =   £0.76/m.

Cost of 150 mm dia. porous concrete pipe=   £8.71/m

Wastage allowance   =   3%

Cost of pipe per metre run   =   1.03 x £8.71   =   £8.97

Cost of granular fill material Type B   =   £11.46/m³

Wastage allowance   =   5%

Depth of trench to invert   =   1.50 m

Thickness of pipe   =   0.013 m

Width of trench   =   0.650 m

Volume of pipe (calculated using the outside diameter of the pipe) per metre run

$$= \quad 3.142 \times (0.176 \times 0.176)/4 \times 1.0$$

$$= \quad 0.024 \text{ m}^3 \text{ per metre run of pipe}$$

Volume of granular fill per metre run of trench

$$= 1.0 \times (1.50 + 0.013) \times 0.65 - 0.024 = 0.96 \text{ m}^3$$

Cost of granular fill material Type B per metre run of trench

$$= \quad 0.96 \text{ m}^3 \times 1.05 \times £11.46 \quad = \quad £11.55$$

Total material cost of pipe per metre run of trench

$$= \quad £0.76 + £8.97 + £11.55$$

$$= \quad £21.28 \text{ per metre run of trench}$$

To this is added the cost of labour and plant which would be calculated in the manner described for storm and foulwater drains.

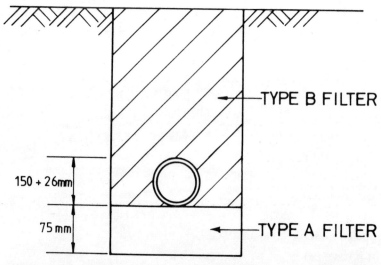

150 + 26mm

75 mm

TYPE B FILTER

TYPE A FILTER

Fig. 5.16 Typical detail for French drain construction

MANHOLES, CATCHPITS AND GULLIES

Because of the complexity of detail in manholes and catchpits these items are time consuming to estimate. Each manhole may have a different depth to invert which makes the quantities involved and hence the resource requirements different in each situation. The estimator may calculate from the drawings the quantities involved for each manhole and then assess the resource requirements and costs to produce a direct cost total. Table 5.22 gives typical quantities required for the construction of a precast concrete manhole 1200 mm in diameter and 1.975 m deep. Table 5.23 gives the prices of the individual resources. Typical usage rates for the operations in the construction of the manhole are given in Tables 5.24 and 5.25. From the figures it is possible to calculate the cost of each individual operation on a unit rate basis.

Table 5.22: Quantities for a precast concrete manhole 1200 mm diameter depth to invert 1.975 m

Excavation (assume square)	6.24 m^3
225 mm deep concrete base slab	0.67 m^3
Precast concrete chamber rings	1.54 m
Step irons	5.00 No.
150 mm concrete surround	1.18 m^3
Shuttering to concrete surround	8.26 m^2
Concrete benching	0.54 m^3
Cover slab	1.00 No.
Brickwork (2 courses, 225 mm)	0.47 m^2
Manhole cover and frame	1.00 No.
(Grade B heavy duty)	
Channel to manhole	1.00 No.

Table 5.23:   Cost of the resources required for the manhole construction

	£
Concrete Grade C20 OPC	49.45/m³
Precast concrete chamber rings	69.05/m
Step irons	4.20/No
Shuttering to the concrete surround	4.00/m²
Precast concrete cover slab	64.94/No.
Brickwork to cover and frame	165.00/thousand
Channel to manhole	4.91/No.
Pipework to manhole	8.61/m
Cast iron cover to manhole	65.32/No.
Labourer	5.10/hour
Ganger	6.69/hour
Tradesman	6.52/hour
Hymac 580 excavator	21.40/hour

Table 5.24:   Operations, gangs and usage rates involved in manhole and catchpit construction

Operation		Gang size	Usages
Eng. brickwork to	112 mm brick	Bricklayer	1.2 gang hr/m²
manhole catchpit	225 mm "	+ labourer	2.4 gang hr/m²
	340 mm "		3.6 gang hr/m²
Build pipes into	150 mm dia.		0.5 man hrs each
manhole walls	225 mm "		0.6 " " "
	300 mm "		0.75 " " "
	375 mm "		1.2 " " "
	450 mm "		1.75 " " "
	525 mm "		2.25 " " "
	600 mm "		3.00 " " "
	675 mm "		3.75 " " "
	750 mm "		4.75 " " "
Fix half round	150 mm		0.1 " " "
channels	225/300 mm		1.3 " " "

Fix half round			
channels	375/675 mm	3.0 man hrs/m²	
Set manhole	light	2.0 " " "	
covers	medium	2.5 " " "	
	heavy	3.0 " " "	
Formwork to	base	2.0 " " "	
manhole	roof slab:		
	sides	1.8 " " "	
	roof soffit	4.8 " " "	
Place concrete in base and roof slabs		2.0 " " "	
Place benching concrete (include formwork)		10.0 man hrs/m³	
Place grano screed - 75 mm thick		1.8 man hrs/m²	
Road gully - exc; set; conc; backfill	Hymac 580 + driver + 2 labourers	2.0 gang hrs/gulley	
Road gully - frame & grating	1 No. bricklayer + 1 No. labourer	2.0 " "	
Fix step iron		0.1 man hrs each	
Off-load, stack, circular precast conc. manholes	22RB+dr+2 labourers per unit	0.1 gang hours	
Uplift, distribute, set in position manholes	22RB+dr+2 labourers/ tractor/trailer + driver	0.5 " "	

Example of excavation and concrete base

Examples are given in Tables 5.26 and 5.27 for the excavation of the manhole and the concreting of the base slab. By adding the cost of each operation required in the construction the total cost of the manhole may be calculated.

The method is essential when a project in an urban environment is being estimated and each manhole needs to be considered individually as a unique structure. For motorways and other roadworks where numerous manholes are involved quicker methods of estimating may be developed.

Table 5.25:   Manhole excavation

Depth (m)	Over-break (%)	Gang hours per m³ as billed	Gang size
0-1.5	75	0.19 gang hours/m³	Hymac 580 + driver + banksman
1.5-3.0	150	0.28 " "	"
3.0-4.5	300	0.55 " "	"
4.5-6.0	500	0.55 " "	22RB back-acter + driver + banksman

Table 5.26:   Example excavation of a manhole

Quantity to be excavated   =   6.24 m³ (bank volume)

Assuming 100% overbreak gives an actual quantity of 12.48 m³

Taking a bulking factor of 1.2 gives a bulked volume of 1.2 x 12.48

=   14.97 m³

Cost of Hymac excavator   @   £ 21.40 per hour

Cost of labourer   @   £   5.10 "   "

A Hymac excavator with a 0.57 m³ bucket may be expected to excavate some 3.5 m³ of material per hour in this work. This means a usage of 0.286 hours per cubic metre of bulked material.

Therefore the labour cost of the excavation =   hours required x quantity x labour rate   =   0.286 x 14.97 x £5.10   =   £21.83.

The plant cost for the excavation =   hours required x plant rate

=   0.286 x 14.97 x £21.40   =   £91.62

Total cost of the excavation for the manhole =   £21.83 +   £91.62

=   £113.45

Table 5.27:   Concreting the base slab of the manhole

Quantity of concrete required   =   0.67 m³

Assume 2.0 man hours per cubic metre to place the concrete.   Concrete is supplied ready mixed   @ £38.50 m³

Cost of labourer   @ £ 5.10 per hour

Material cost of base slab   = 0.67 x £38.50   =   £25.79

Cost of placing the concrete = number of hours x quantity x labourer

rate

= 2.0 x 0.67 x £5.10   =   £6.83

Total cost of concreting base = £25.79 + £6.83 =   £32.62

Table 5.28:   Quantities for precast concrete manholes
1.5 m deep to invert

Operation or item		Ring diameter					
	mm	914	1067	1219	1372	1524	1829
	in.	36	42	48	54	60	72
Excavate (square)	m³	3.15	3.90	4.86	5.92	6.93	9.37
Conc. base (150 mm)	m³	0.29	0.35	0.45	0.54	0.63	0.85
Chamber rings	m	1.07	1.07	1.07	1.07	1.07	1.07
Step irons	No.	3	3	3	3	3	3
Conc. surrounds (150 mm)	m³	0.60	0.70	0.80	0.86	0.95	1.13
Shuttering	m²	4.53	5.10	5.70	6.29	6.81	7.91
Conc. benching	m³	0.20	0.28	0.54	0.90	1.39	2.40
Rendering	m²	0.67	0.92	1.17	1.50	1.84	2.68
Pipes	Item	Item	Item	Item	Item	Item	Item
Cover slab	No.	1	1	1	1	1	1
Brickwork to slab (225 mm, 2 courses)	m²	0.47	0.47	0.47	0.47	0.47	0.47
Cover & frame	No.	1	1	1	1	1	1
Channels	Item	Item	Item	Item	Item	Item	Item

Precast concrete manholes

Where it is necessary to estimate standard sized precast concrete manholes a short cut method of utilising the data shown in Tables 5.28 and 5.29 may be used.   Table 5.28 contains the material quantities required for precast concrete manholes of 1.5 m depth to invert. Included in this list of quantities are materials such as the channel and cover slab, the quantities of which are independent of any particular depth of manhole.   Table 5.29 contains the material quantities required for each 100 mm variation in the depth of the manhole.   Data are given for a shaft ring and main chamber ring.   By using an appropriate factor in these data to allow for the depth of manhole over and above 1.5 m the quantities required may be calculated.

Example of estimating for precast manhole

Quantities

It is required to calculate the quantities in a precast concrete manhole 1219 mm in diameter with a depth to the invert of 2.1 m.   Table 5.28 is referred to in order to obtain the quantities for the depth to 1.5 m.   This leaves the quantities for an extra $2.1 - 1.5 = 0.6$ metres to be calculated from the data in Table 5.29.   This is summarised in Table 5.30.

Material cost

Using the total quantities from Table 5.30 and the resource prices from Table 5.23 the material cost of the manhole is calculated as in Table 5.31.

Labour and plant cost

The labour and plant requirements are calculated on an operational rate basis by calculating the weekly cost of a gang and then dividing by the anticipated number of completed manholes per week to obtain a unit labour and plant cost.   An example of the calculation of the labour and plant costs is given in Table 5.32.

Table 5.29: Precast concrete manhole quantities variations per 100 mm

Operation or item		Ring diameter					
	mm	914	1067	1219	1372	1524	1829
	in.	36	42	48	54	60	72
CHAMBER RING							
Excavate	m^3	0.19	0.24	0.29	0.36	0.42	0.57
Chamber rings	m	0.10	0.10	0.10	0.10	0.10	0.10
Step irons	No.	0.33	0.33	0.33	0.33	0.33	0.33
Conc. surrounds	m^3	0.06	0.07	0.08	0.08	0.09	0.11
Shuttering	m^2	0.43	0.49	0.54	0.60	0.65	0.75
SHAFT RING							
Excavate	m^3	0.19	0.24	0.29	0.36	0.42	0.57
686 Shaft ring	m	0.10	0.10	0.10	0.10	0.10	0.10
Step irons	No.	0.33	0.33	0.33	0.33	0.33	0.33
Conc. surround	m^3	0.05	0.05	0.05	0.05	0.05	0.05
Shuttering	m^2	0.35	0.35	0.35	0.35	0.35	0.35

Table 5.30: Calculation of quantities for a precast concrete manhole 1219 mm dia., 2.1 m deep

Operation or item	Units qty	A 1.5 m	B 0.1 m	C (Bx6)	Total (A+C)
Excavate (square)	m^3	4.86	0.29	1.74	6.60
Concrete base					
(150 mm)	m^3	0.45	-	-	0.45
Chamber rings	m	1.07	0.10	0.60	1.67
Step irons	No.	3	0.33	2	5
Conc. surround					
(150 mm)	m^3	1.28	0.80	0.08	0.48
Shuttering to surround	m^2	5.70	0.54	3.24	8.94
Conc. benching	m^3	0.54	-	-	0.54

Rendering to benching	m²	1.17	-	-	1.17
Pipes	Item	Item	-	-	Item
Cover slab	No	1	-	-	1
Brickwork to slab					
(225 mm, 2 courses)	m²	0.47	-	-	0.47
Manhole cover & frame	No.	1	-	-	1
Channels	Item	Item	-	-	Item

Total cost of manhole

Total cost = material cost + labour and plant cost

= £425.78 + £355.65 = £781.43 each.

Table 5.31: Calculating the material cost of the manhole

Concrete to base	0.45	m³	@ £ 38.50	=	£	17.32
Precast concrete chamber rings	1.67	m	@ £ 69.05	=	£	115.31
Step irons	5	No.	@ £ 4.20	=	£	21.00
Concrete surrounds	1.28	m³	@ £ 38.50	=	£	49.28
Shuttering to the concrete surround	8.94	m²	@ £ 4.00	=	£	35.76
Concrete benching	0.54	m³	@ £ 38.50	=	£	20.79
Rendering to benching	1.17	m²	@ £ 7.70	=	£	9.01
Pipework	2.00	m	@ £ 8.61	=	£	17.22
Cover slab	1	No.	@ £ 64.94	=	£	64.94
Brickwork	0.47	m²	@ £ 10.46	=	£	4.92
Manhole cover & frame	1	No.	@ £ 65.32	=	£	65.32
Channels	1	Item	@ £ 4.91	=	£	4.91
		Total material cost		=	£	425.78

Catchpits

Catchpits are similar in construction to manholes but are of a more simple design. The designs are also more standard.

Gullies

The construction of gullies and gully connections is normally estimated on a unit rate basis as in the example in Table 5.33.

CONNECTIONS TO BOTH NEW AND EXISTING DRAINAGE

Where connections have to be made to new or existing pipework the drawings must be carefully studied to ensure all available information is obtained. The work involved should be listed as a series of operations each of which is then estimated on a unit rate basis.

Where the work is of a more repetitive nature (e.g. on motorway drainage construction) then the connections may be estimated on an operational basis. For larger quantities of pipe connection the same method of estimating may be used as for the storm and foulwater drainage calculations but with larger output rates assumed for the gang and a higher wastage percentage allowed on the pipework quantities.

Table 5.32: Calculation of the labour and plant costs for the manhole

---

The gang requirements for the construction of the manholes are as follows:

Labour:	1 ganger	@ £6.69/hour	= £ 6.69
	3 labourers	@ £5.10/hour	= £ 15.30
	1 craftsman	@ £6.30/hour	= £ 6.30
Labour cost per hour			= £ 28.29

Plant:	1 Hymac excavator	@ £21.40/hour	= £ 21.40
	1 tractor & trailer	@ £10.25/hour	= £ 10.25
	1 dumper	@ £11.19/hour	= £ 11.19
Plant cost per hour			= £ 42.84

Total labour and plant cost per hour   =   £28.29 + £42.84 = £71.13

Assuming a 50-hour week, cost per week =   50 x £71.13 = £3556.50

It is anticipated that 10 manholes will be completed each week therefore

unit labour and plant cost   =   £3556.50/10   =   £355.65 each

---

Table 5.33:   The calculation of the cost of a gully pot

Labour and Plant

A typical labour and plant gang for the construction of gullies would consist of:

1 ganger	@	£ 6.69/hour	
2 labourers	@	£5.10/hour each	= £10.20/hour
1 JCB 3CX	@	£ 19.90/hour	

Total gang cost   =   £6.69 + £10.20 + £19.90   =   £36.79/hour

Assume 50-hour week   =   weekly cost of 50 x £36.79   =   £1839.50

The gang should complete 25 gullies per week giving a unit cost of £1839.50/25   =   £73.58 per gully

Materials Cost

Gully pot and grating	=	£61.00 each
Brickwork to gully	=	£16.50 each
Concrete surround	=	0.6 m³ @ £38.50/m³ = £23.10
Additional pipework for connection	=	1.0 m @ £8.61/m = £8.61.
Total material cost	=	£61.00 + £16.50+ £23.10 + £8.61 = £109.21.
Total cost	=	labour and plant + material
	=	£73.58 + £109.21 = £182.79

DRAINAGE TO THE TOE OF EMBANKMENTS AND TOP OF CUTTINGS

Drainage channels, ditches and pipeworks to the toe of embankments and at the top of the cut areas must be completed at an early stage in the contract before the start of the main earthworks programme.

The various lengths of drainage works are abstracted, totalled and suitable rates of production assumed. For example: 'V' ditching as shown in Fig. 5.17 will require a JCB 3 and 1 labourer.

Cost of JCB 3   =   £ 19.90/hour

Cost of labourer   =   £   5.10/hour

Total cost of gang   =   £19.90 + £5.10   =   £25.00/hour

Assume an output rate of 5 m per hour, then for a 175 m length of ditch total cost   =   175/5 x £25.00   =   £875.00

Cost per metre run = £875.00/175   =   £5.00

Where the ditch is to be concrete lined or random rubble lined additional labour and material allowance must be made.

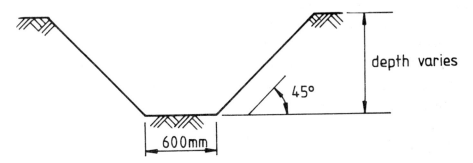

Fig. 5.17   Typical cross section of a 'V' ditch

PIPED CULVERTS

Where existing water courses cross the line of the road it is necessary to construct culverts before it is possible for the earthworks in that area to commence.   Depending on the volume of water to be accommodated the culvert may take the form of a piped culvert or a concrete box culvert.   The most usual form is that of the piped culvert.

Drainage pipes of greater than 600 mm diameter require additional labour and plant for handling of pipes and materials over and above the normal sized drainage gang.   Where a concrete surround to the pipe has been specified by the client's advisors the details may require formwork and reinforcement.

The culvert must be treated as an individual structure.   In addition to the actual construction, provision must be made to:
-   maintain the flow of the existing watercourse;
-   block off existing watercourse at either end of the culvert;
-   excavate and dispose of unsuitable material;   and
-   import and compact granular fill material.

A typical labour and plant gang for the construction of a piped culvert will consist of:

1 ganger and 4 labourers;

1 Hymac excavator;

1 RB22 tracked crane;

1 JCB 3CX excavator;   and

1 Bomag 90 roller.

Each culvert must be planned as a separate section of the work and the resource requirements must be analysed to ensure the labour and plant resources are fully integrated with the requirements of the main structures.

FORMWORK

Introduction

The basic requirements for formwork for the construction of concrete structures are:

* to support the fresh concrete in the required level and alignment until it is self supporting or until the structure produced is capable of carrying any imposed loads in addition to its self-weight;

* to form the required structures in as economical a manner as possible consistent with the requirements of the contract;   and

* to impart any specified finish to the concrete surface.

Common materials used as facing for formwork are:

* timber;

* plywood;

* steel;

* glass fibre reinforced plastics;

* P.V.C.;

* expanded polystyrene;   and

* expanded metal.

The formwork designer may select one or a combination of several materials for the construction of a particular structure. The main criteria in selection of the materials are:

* the number of re-uses required;

* the surface finish specified by the engineer;

* the tolerances specified;

* weight considerations;
* whether the formwork is a proprietary make or purpose built;
* the skilled manpower available for making, erecting and striking the formwork;   and
* the availability of materials.

From the selected type of material and the design of the formwork the estimator must calculate the material and labour cost required to make, erect and then strike the formwork. Suitable allowances must be made for wastage of materials, sundry items and formwork shutter treatments.

Materials and labour

Formwork is usually estimated on a unit rate basis. Tables 5.34 to 5.38 show typical material and labour requirements per square metre for formwork to:
* foundations;
* walls;
* soffits;   and
* columns and beams.

Tables 5.39 and 5.40 show typical requirements for other general formwork items.

Table 5.34:   Formwork for foundations

Description	Materials per m^2		Hours per m^{2*}	
	Type	Quantity	Make	Fix and strike
COLUMN BASES & PILE CAPS				
Under 0.5 m^3 vol	Timber	0.04 m^3	1.10	1.10
	Plywood	1.00 m^2		
0.5 to 1.5 m^3 vol	Timber	0.05 m^3	0.90	1.05
	Plywood	1.00 m^2		
Over 1.5 m^3 vol	Timber	0.07 m^3	0.80	1.00
	Plywood	1.00 m^2		
GROUND BEAMS	Timber	0.06 m^3	0.90	1.10

GROUND BEAMS	Plywood	1.00 m²		
GROUND SLAB EDGES	Timber	0.05 m³	1.35	2.00
150 mm high or less	Plywood	1.10 m²		
GROUND SLAB EDGES	Timber	0.06 m³	0.90	1.10
exceeding 150 mm high	Plywood	1.00 m²		
STRIP FOOTINGS	Timber	0.04 m³	0.60	0.80
	Plywood	1.00 m²		
COVER SLABS TO	Timber	0.07 m³	1.50	2.00
MANHOLES OR DUCTS	Plywood	1.00 m²		

Stop ends through light reinforcement 2 hours per m²

Stop ends through heavy reinforcement 3 hours per m²

* Note: The hours for making are for making once and are to be divided by the number of uses. The hours per m² are at composite joiner rate.

Table 5.35:   Formwork for walls

Material	Materials per m²		Hours per m²*	
	Type	Quantity	Make strike	Fix &
WALLS				
One side shuttered	Timber	0.04 m³	0.90	1.20
	Plywood	1.00 m²		
Both sides shuttered	Timber	0.40 m³	0.90	1.10
	Plywood	1.00 m²		
DUCT WALLS				
One side shuttered	Timber	0.05 m³	1.00	1.40
	Plywood	1.00 m²		
Both sides shuttered	Timber	0.04 m³	1.00	1.30
	Plywood	1.00 m²		
Walls using Acrow	Pans	1.00 m²	-	0.76
formwork	Timber	0.02 m³		

* Note: The hours for making are for making once and are to be divided by the number of uses. The hours per m² are at composite jointer rate.

Table 5.36: Additional items to walls (quantities of materials to depend on actual requirements)

Description	Unit	Hours/ Unit M+F+S*
Wall kickers fixed direct	Metre of wall	0.40
Ditto hanging	Metre of wall	0.80
Stop ends n.e. 300 mm thick	Linear metre	0.90
Ditto exceeding 300 mm thick	m²	3.00
Form openings n.e.150 x 150 x 150 mm thick No.		0.38
" " " " 300 x 300 x 150 mm thick No.		0.60
" " " " 150 x 150 x 300 mm thick No.		0.50
" " " " 300 x 300 x 300 mm thick No.		0.80

* Note: M+F+S denotes Make, Fix and Strike

Table 5.37: Formwork for soffits

Description	Materials per m²		Hours per m²'	
	Type	Quantity	Make strike	Fix &
All timber (boards)	Timber	0.09 m³	2.70	1.30
Joists & bearers with	Timber	0.02 m³		
plywood	Plywood	1.00 m²	0.50	1.30
Floor centres with plywood	Timber	0.01 m³		
	Plywood	1.00 m²	0.40	1.30
Steel pans & floor centres	Timber	0.01 m³		
	Steel pans	1.00 m²	0.30	1.30
Trolley system using crane	Depends on design		-	0.40
Fibreglass moulds	Depends on design		-	0.75
Trough floors	Depends on design		-	1.00
Staircases	Timber	0.06 m³		
	Plywood	1.00 m²	2.00	2.00

Temporary supports to

permanent formwork   Timber   0.02 m³ 0.50   0.50

Unload and handle :   Steelpans   0.3 hours per m²
                                     Fibreglass
                                     moulds   0.2   "   "
                                     Trough floors   0.2   "   "

For small areas increase the soffit rate as follows:

   Area   41-50 m² + 25%

    "   25-40 m² + 33%

    "   10-24 m² + 50%

    "   0-9 m² + 100%

Areas are per contract and not individual areas.

For example: if a contract has several areas of, say, 25 m², then the standard rates should be used.

For slab thickness requiring closer bearing centres and increased support, the following additions should be used.

        Slab thickness over   250 - 300 mm + 5%

           "   "   "   300 - 350 mm + 15%

           "   "   "   350 - 400 mm + 25%

           "   "   "   400 - 450 mm + 35%

           "   "   "   450 mm   + 45%

For extra high soffits the following additions should be used.

        Slab height not exceeding   3.5 m basic

                            3.5 - 5.0 m high + 25%

                            5.0 - 6.5 m high + 40%

---

* Note: The hours for making are for making once and are to be divided by the number of uses.

         The hours per m² are at composite joiner rate.

Table 5.38:   Formwork for columns and beams

Description	Materials per m²		Hours per m²	
	Type	Quantity	Make	Fix & strike
COLUMNS & STANCHIONS				
Average size 300 x 300 mm	Timber	0.06 m³	1.20	1.20
	Plywood	1.00 m²		
" " 375 x 375 mm	Timber	0.05 m³	1.10	1.10
	Plywood	1.00 m²		
" " 40 x 450 mm	Timber	0.05 m³	1.00	1.10
	Plywood	1.00 m²		
" " 525 x 25 mm	Timber	0.04 m³	0.90	1.00
	Plywood	1.00 m²		
" " 600 x 600 mm	Timber	0.04 m³	0.90	0.90
	Plywood	1.00 m²		
BEAMS				
Under 0.1 m² sectional area	Timber	0.08 m³	1.30	1.60
" "	Plywood	1.00 m²		
0.1 to 0.15 m² " "	Timber	0.07 m³	1.20	1.50
	Plywood	1.00 m²		
Over 0.15 m² " "	Timber	0.06 m³	1.10	1.40
	Plywood	1.00 m²		
EDGE BEAMS				
Under 0.1 m² sectional area	Timber	0.06 m³	1.30	1.90
	Plywood	1.00 m²		
Over 0.1 m² " "	Timber	0.05 m³	1.10	1.70
	Plywood	1.00 m²		

Notes:

If beam clamps are used reduce timber by 0.02 m³ per m². Allow 0.6 hours per column kicker.

For beams not exceeding 2 m long add 50% to rate.

* Note: The hours for making are for making once and are to be divided by the number of uses. The hours per m² are at composite joiner rate.

Table 5.39:   General formwork items per m²

Description	Materials per m²		Hours per m²*	
	Type	Quantity	Make	Fix & strike
CABLE TRENCHES, ducts etc.	Timber	0.07 m³	2.70	1.30
in thickness of concrete bed	Plywood	1.00 m²		
CANOPIES & LANDINGS	Timber	0.06 m³	2.00	2.50
	Plywood	1.00 m²		
PROJECTING SOFFITS	Timber	0.12 m³	2.00	2.20
of slab up to 450 mm wide	Plywood	1.00 m²		
Ditto up to 600 mm wide	Timber	0.11 m³	2.00	2.20
	Plywood	1.00 m²		
Ditto up to 750 mm wide	Timber	0.09 m³	1.70	2.20
	Plywood	1.00 m²		
STAIRS, STRINGERS &	Timber	0.07 m³	3.00	2.00
RISERS	Plywood	1.00 m²		
CONCRETE				
CIRCULAR ON PLAN				
Columns average 300 mm dia.	Timber	0.05 m³	5.40	1.50
	Hardboard	1.00 m²		
	Plywood	1.40 m²		
Ditto average 600 mm dia.	Timber	0.04 m³	4.90	1.35
	Hardboard	1.00 m²		
	Plywood	1.30 m²		
Ditto average 900 mm dia.	Timber	0.04 m³	4.30	1.25
	Hardboard	1.00 m²		
	Plywood	1.20 m²		
WALLS CIRCULAR ON PLAN	Timber	0.10 m³	3.00	2.20
	Plywood	1.00 m²		

* Note: The hours for making are for making once and are to be divided by the number of uses. The hours per m² are at composite joiner rate.

Table 5.40:   General formwork per linear metre

| Description | Materials per m | | Hours per m* | |
	Type	Quantity	Make	Fix & strike
Hours per linear metre - composite joiner rate				
SUSPENDED SLAB EDGES	Timber	0.02 m³	0.30	0.40
not exceeding 150 mm high	Plywood	0.15 m²		
Ditto over 150 mm high	Timber	0.025 m³	0.40	0.50
	Plywood	0.25 m²		
FORM CONSTRUCTION JOINT				
Rates as slab edges	-	-	-	-
EXTRA OVER FOR SPLITTING				
OVER REINFORCEMENT	-	-	-	0.35
Temporary turning piece and				
strutting to soffit of flat arch				
113 mm wide	Timber	0.01 m³	0.30	0.50
Ditto 225 mm wide	Timber	0.02 m³	0.50	1.00
Hours per number - composite joiner rate				
FORM OPENINGS IN WALLS				
USING POLYSTYRENE				
n.e. 150 x 150 x 150 mm	Polystyrene	0.004 m³	-	0.16
n.e. 300 x 300 x 150 mm	"	0.015 m³	-	0.26
n.e. 150 x 150 x 300 mm	"	0.007 m³	-	0.25
n.e. 300 x 300 x 300 mm	"	0.030 m³	-	0.38

* Note: The hours are for making once and are to be divided by the number
of uses. The hours per m are at composite joiner rate.

Wastage of materials

Appropriate allowance of, say, 10% to 20% must be made for wastage during the making of the formwork. The following are additional allowances made for waste due to reuse.

Up	to	4	uses	Nil
5	-	6	"	10%
7	-	8	"	12.5%
		9	"	15%
10	-	11	"	17.5%
		12	"	20%

An extra material allowance of, say, 5% should be made for formwork to battered walls or other special features.

Sundry items

The estimator must make an allowance for all the sundry items (props, ties, loops, clamps, etc.) that will be required in addition to the formwork material to keep the formwork in place until struck. Some of these items will be re-usable, and their cost may be apportioned over a number of uses of the formwork. Other items may be cast in the concrete and the total must be included within the bill item rate.

Formwork shutter treatments

The faces of the formwork which will be in contact with the concrete must be treated with a suitable release agent to prevent the concrete adhering to the formwork and causing damage to the final surface of the concrete when the formwork is struck. The estimator must allow for both the cost of the material and the time spent to treat the shuttering. Different allowances for different types of formwork shutter treatments are shown in Table 5.41.

Table 5.41:   Formwork shutter treatments

| Operation | Materials per m² | | Hr/m² |
	Type	Quantity	
Stop up screw holes & blemishes	Cement & Multibond	Nom.	0.06
Ordinary oiling	Oil	0.36	0.04
Standard Moldcote (Note: applied in 3 coats, oiling between coats),			
rate per coat	Moldcote	0.06	0.11
Ditto without oiling 1st coat	Moldcote Primer	0.10	0.12
Remaining 2 coats	Moldcote	0.21	0.22
Moldcote applied 1-4 coats per coat		0.06	0.11
Joint sealer	Sealing tape	incl	incl
Mould oil substitute			
in one coat : to wood		0.11	0.11
: to metal		0.10	0.10
Mould & shutter			
sealer applied in 2 coats (oiling			
in between) rate per coat		0.07	0.11
Enfield mould coating applied			
in 2 coats (oiling in between),			
rate per coat	697	0.10	0.11
Polyurethane paint applied in 2 coats			
(oiling in between), rate per coat		0.09	0.11
Wire brush and clean down	-	-	0.22
Treat with retarding fluid,			
to form key	Fluid	0.18	0.09

Make good any of the above: Add 10% per square metre

Example of calculation of cost of standard formwork panel

Wherever possible formwork will be constructed using standard panels of size 2.44 m x 1.22 m (8 ft x 4 ft). These may be manhandled into position and combined for various different uses. The following calculation shows the cost estimate for a typical panel shown in Fig. 5.18. The calculation assumes ten uses of the panel.

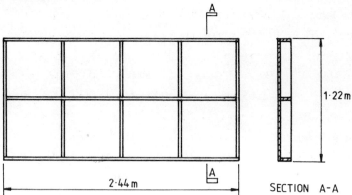

PANEL CONSTRUCTION : 18 mm PLYWOOD FACING BACKED BY 100 mm × 50 mm SOFTWOOD STRUTTING.

Fig. 5.18   Typical standard formwork panel

Cost of material

Volume of timber required	=	(3 x 2.44 x 0.1 x 0.05)
		+ (5 x 1.22 x 0.1 x 0.05)
	=	0.04 + 0.03 = 0.07 m³

18 mm plywood facing

Area of ply	=	2.44 x 1.22 = 2.98 say 3.0 m²		
Cost of timber	=	£253.40/m³		
Cost of 18 mm ply	=	£6.59/m²		
Allow 17.5% wastage				
Therefore cost of timber	=	0.07 x 1.175 x 253.40	=	£ 20.84
Cost of ply	=	3.00 x 1.175 x 6.59	=	£ 23.23
		Total	=	£ 44.07
Allow for nails at £1.66/m²			=	£ 4.98
		Total	=	£ 49.05

Total material cost	= £ 49.05
Cost of 1 use of panel	= £ 49.05
Cost per 1 use per m²	= £ 16.47
Cost per m² 10 uses	= £ 1.65

Labour cost to make panel

Assume gang of 2 carpenters and 1 labourer.

Carpenter cost	= £6.30 per hour
Labourer cost	= £5.10 per hour
Composite labour rate	= (2 x £6.30 + £5.10)/3
	= £5.90/hour
Labour requirement per m² of panel	= 1.25 hours
Total labour cost = 1.25 x 3 x £5.90	= £ 22.12
Labour cost for 1 use of panel	= £ 22.12
Cost of labour per m² of panel	= £ 7.43
Cost of labour per m² of panel 10 uses	= £ 0.75

Labour cost to fix and strike panel

Assume 1.55 labour hours per m² to fix and strike panel.

Cost = 1.55 x £5.90 = £9.14/m²

Clean and oil panel

This operation will require 0.04 hours labour per square metre. A suitable mould oil would cost £1.80 per litre and 0.36 litres would be required per square metre of panel. Therefore the cost to clean and oil the panel

$$= (0.04 \times 5.90) + (0.36 \times 1.80) = 0.24 + 0.65$$
$$= £0.89/m^2$$

Allowance for props, bolts and ties = £2.80/m²

Total cost

Total cost to make, fix and strike panel assuming 10 uses.

= 1.65 + 0.75 + 9.14 + 0.89 + 2.80 = £15.23/m²

Example of calculation of formwork for a retaining wall

Assume the shutter is to be constructed of eight sheets of plywood each 2.44 m x 1.22 m backed by 100 mm x 75 mm timber at 600 mm centres both horizontally and vertically as shown in Fig. 5.19. It is assumed that ten uses will be made of the shutter.

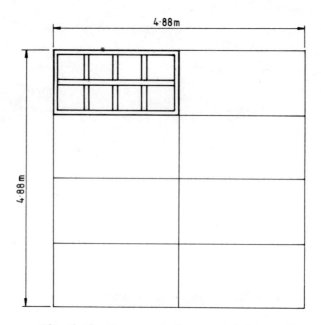

Fig. 5.19   Formwork for a retaining wall

Cost of material

Plywood	-	Total area = 8 x 1.22 x 2.44	=	23.81 m²
		Assume 17.5% wastage		
		Cost of 18 mm ply	=	£6.59/m²
		Total cost of ply	=	23.81 x 1.175 x 6.59
			=	£184.37
Timber	-	Total quantity of timber	=	(8 x 3 x 2.44 x 0.1 x 0.075)
				+ (8 x 5 x 1.22 x 0.1 x 0.075)
			=	0.439 + 0.366 = 0.805
		Assume 17.5% wastage		
		Cost of timber	=	£253.40/m³

Total cost of timber   =   0.805 x 1.175 x 253.40

          =   £239.68

Cost of bolts, nails and screws, say £3.50 per sheet of ply.

Therefore total material cost   =   £184.37 + 239.68 + 28.00

          =   £ 452.05

Cost of 1 use of the formwork   =   £ 452.05

Cost of 1 use per m²   =   £ 18.99

Cost per m² 10 uses   =   £ 1.90

Labour cost to make formwork

Assume gang of   1 chargehand carpenter @ £6.90 per hour

         2 carpenters @ £6.30 per hour

         1 labourer @ £5.10 per hour

Average labour cost = (6.90 + 12.60 + 5.10)/4 = £6.15 per hour

Allow 1.25 labour hours to make a square metre of the shutter.

Total labour cost = 1.25 x 23.81 x 6.15 = £183.04

Cost of labour for 1 use of the formwork = £183.04

Cost of labour per square metre   =   £ 7.69

Cost of labour per square metre 10 uses   =   £ 0.76

Cost to fix and strike the panel

Allow 1.62 labour hours per square metre to fix and strike.

Labour cost per square metre = 1.62 x 6.15 = £9.96

Clean and oil panel

Allow 0.04 hours labour per metre to clean and oil.

Cost of mould oil per square metre =   £0.65

Total cost = (0.04 x 5.10) + 0.65 =   £0.85/m²

Support to formwork

The support to the formwork is calculated on an operational basis. Assume a total time period of six weeks.

Support requirements:

10 soldiers (3.429 m)	@	£1.40 each	=	£	14.00
10 soldiers (0.889 m)	@	£0.50 "	=	£	5.00
10 Channel splices	@	£0.24 "	=	£	2.40
4 Plumbing frames	@	£0.40 "	=	£	1.60
4 Tilting props	@	£0.84 "	=	£	3.36
25 m lengths tubing	@	£0.80 per m	=	£	2.00
18 Double couplers	@	£0.07 each	=	£	1.26
18 Swivel couplers	@	£0.07 "	=	£	1.26
Total cost per week			=	£	30.88

Total cost = £30.88 x 6	=	£185.28	
Assume 10 uses; then cost per use	=	£ 18.53	
Cost/m² assuming 10 uses	=	£ 18.53 / 23.81	
	=	£ 0.78 per m²	
Assume 20% loss of materials	=	£ 0.16 per m²	
Then total cost of support structure	=	£ 0.94 per m²	

Assume coil ties @ 300 mm centres - total required 34

Cost of ties 34 x £0.80	=	£ 27.20
Assume 10% wastage 1.10 x £27.20	=	£ 29.92
Cost per m² = £29.92/23.81	=	£ 1.26
Tie bolts & workers 24 sets @ £1.80 each	=	£ 43.20
Assume 10 uses	=	£ 4.32 per use
Cost per m²	=	£ 0.18

Total cost

Total cost to make, fix and strike the formwork (assuming 10 uses)
= 1.90 + 0.76 + 9.96 + 0.85 + 0.94 + 1.26 + 0.18
= £15.85 per square metre.

REINFORCEMENT

Reinforcement for concrete structures may consist of round steel bars, twisted or deformed bars and interwoven or welded mesh fabric. The following British Standard Specifications have been produced:

BS 4449: Specification for hot rolled steel bars for the reinforcement of concrete (1981).

BS 4461: Specification for cold worked steel bars for the reinforcement of concrete (1981).

BS 4483: Steel fabric for concrete reinforcement (1969).

BS 4466: Bending dimensions and scheduling of reinforcement for concrete (1981).

Bar reinforcement

The diameter of the reinforcement bar determines the weight per metre length and cross-sectional area.   Table 5.42 shows the size and weight of bar reinforcement.

Table 5.42:   Diameters, weights and areas of reinforcement bars

Bar diameter (mm)	Weight (kg/m)	Sectional area (mm^2)
6	0.222	28.3
8	0.395	50.3
10	0.616	78.5
12	0.888	113.1
16	1.579	201.1
20	2.466	314.2
25	3.854	490.9
32	6.313	804.2
40	9.864	1256.6

Wastage allowances

The bars may be purchased as random lengths, cut lengths or cut and bent to the contract schedules. The form in which the reinforcement is delivered to the site determines the allowance that must be made for wastage. The following is a typical example:

Steel purchased in random lengths   10%

Steel purchased in cut lengths   5%

Steel purchased cut and bent   2.5%

Handling allowances

On delivery the reinforcement must be off-loaded and transported to the actual location in which it is required. This task may be performed by hand or use of cranage and a tractor and trailer. Typical values for unloading and transportation are shown in Table 5.43.

Table 5.43:   Handling allowances for reinforcement bars (hours per tonne)

Operation	Bar diameters (mm)								
	6	8	10	12	16	20	25	32	40
Unload steel									
By hand	2	2	2	2	2	2	2	2	2
By crane	0.5	0.5	0.5	0.5	0.5	0.5	0.5	0.5	0.5
Transport steel by hand									
n.e. 60 m	3	3	3	3	3	3	3	3	3
n.e. 100 m	4	4	4	4	4	4	4	4	4
n.e. 120 m	5	5	5	5	5	5	5	5	5
By trailer including load and unload									
By hand	4	4	4	4	4	4	4	4	4
By crane	1	1	1	1	1	1	1	1	1

Steel fixer

The steel fixer may be required to undertake all or part of the following operations:

* cut the steel by hand or machine;
* bend the steel by hand or machine; and
* fix the steel.

Some allowance may also have to be made for lifting the reinforcement into position by crane or hoist.   Table 5.44 shows typical hours per tonne required for steel fixer operations.

Table 5.44:   Steel fixer allowance: for cutting, bending and hoisting reinforcement bars (hours per tonne)

Operation	Bar diameters (mm)								
	6	8	10	12	16	20	25	32	40
Hand cut	9	8	7	6	5	4	-	-	-
Hand bent	33	26	21	17	14	11	-	-	-
Machine cut	7	6	5	4	4	3	3	2	2
Machine bent	26	20	16	14	10	8	6	5	4
Fixing steel including the use of a mechanical hoist									
	60	42	32	25	18	14	13	11	9
Extra over for hoisting by hand									
n.e. 15 m	4	4	4	4	4	4	4	4	4
over 15 m	7	7	7	7	7	7	7	7	7

Additional allowances

The estimator must make allowance for all additional items relating to the steel fixing.   These include provisions for:
* the cutting of mild steel bars by acetylene;
* binding wire;   and
* spacer blocks and chairs.

For small quantities of links, stirrups and chairs an addition of 25% may be made to the hours for cutting, bending and fixing.   The provision of chairs to support the top reinforcement is frequently allowed for in the wastage percentage of the bars.

Mild steel reinforcement may be cut by acetylene and an allowance of 4 hours per tonne is usually made.

Typical quantities of binding wire per tonne of reinforcement are shown in Table 5.45.

Table 5.45:   Binding wire

Bar diameter (mm)	Kg/tonne
6	18
8	18
10	13
12	11
16	9
20	7
25 and over	5

The spacers and chairs required for a contract depend on the system to be used. In general a typical allowance of 1.5% to the steel price would be made for horizontal elements and for vertical elements 100% should be added to the binding wire price. The labour hours required to fix spacer blocks are normally allowed for in the overall fixing rate.

Even if the reinforcement is being delivered to the site cut and bent it is usual to include within the estimate an allowance for providing cutting and bending on site to ensure that any additional bending of bars may be made if required.

The calculation of weighted average rates

Under the Civil Engineering Standard Method of Measurement (CESMM2) each different diameter of bar reinforcement should be measured separately and listed as a separate bill item. The estimator will calculate an overall rate for the various diameters of reinforcement and apply it accordingly to the relevant items.

In bills of quantities drawn up under the Method of Measurement for Highway Works, bar reinforcement is measured as that of nominal size 16 mm and under and that which is of nominal size 20 mm and over. The estimator will normally request from the client's advisors the bar bending schedules for the contract and calculate the tonnage of each different bar diameter. This enables a weighted average rate to be calculated and applied against the relevant item.   An example of the calculation is shown in Table 5.46.

**Table 5.46: Calculation of weighted average rate for
16 mm diameter bars and under**

Size	Weight (tonnes)	Rate (tonne)	Total cost
10 mm dia.	8.400	£437.45	£3674.58
12 mm dia.	12.502	£438.11	£5477.25
16 mm dia.	31.671	£405.11	£12830.24
Total weight	52.573		£21982.07

Weighted average rate 16 mm diameter bars and under
= £21982.07/52.573 = £418.12

A similar calculation is performed for bars of 20 mm diameter and over.

Mesh reinforcement

Mesh reinforcement is used as nominal reinforcement for concrete slabs or as distribution reinforcement and is measured in square metres.   The weight in kg/m^2 is dependent on the type and size of the mesh.   Table 5.47 summarises the sizes and weights of mesh produced to BS 4483.

Table 5.47: Mesh types and sizes to BS 4483

Ref. No.	Weight (kg/m²)	Fabric type	
A 393	6.16	Square mesh	200 x 200
A 252	3.95	Square mesh	200 x 200
A 193	3.02	Square mesh	200 x 200
A 142	2.22	Square mesh	200 x 200
A 98	1.54	Square mesh	200 x 200
B 1131	10.90	Structural	100 x 200
B 785	8.14	Structural	100 x 200

B 503	5.93	Structural	100 x 200
B 385	4.53	Structural	100 x 200
B 283	3.73	Structural	100 x 200
B 196	3.05	Structural	100 x 200
C 785	6.72	Long mesh	100 x 400
C 503	4.34	Long mesh	100 x 400
C 385	3.41	Long mesh	100 x 400
C 283	2.61	Long mesh	100 x 400
D 49	0.766	Wrapping	100 x 100
D 31	0.493	Wrapping	100 x 100

A general allowance of 15% is made above the measured quantity to allow for wastage and the necessary overlapping of sheets. Mesh fabric is usually delivered in standard sheets size 4.8 m x 2.4 m. Allowance should be made for the cutting of the sheets to suit the requirements of the structure.

Typical calculations for an overbridge structure

Pepperbox Farm Bridge

Mild steel rebar less than or equal to 16 mm dia. = 14.0 tonnes. Mild steel rebar equal to or greater than 20 mm dia. = 23.0 tonnes. From the bar bending schedules supplied by the client's advisors the weight of each bar diameter is calculated. The figures are shown in Table 5.48.

Calculation of the labour and material rates for 12 mm diameter reinforcement is as follows.

Labour rate
The cost of the steel fixer = £6.30 per hour
Steel fixer hours per tonne of reinforcement for 12 mm diameter rebar (supplied cut and bent):
Off load   :   2 hours per tonne
Transport :   4 hours per tonne

Fix   :   25 hours per tonne

Total steel fixer requirement = 2 + 4 + 25 = 31 hours per tonne

Labour rate per tonne at a cost of £6.30 per hour = 31 x £6.30 = £195.30

Material rate

12 mm diameter rebar (supplied cut and bent) = £438.11 per tonne; allow 2.5% to cover wastage, binding wire, etc. Then the rate for 12 mm diameter rebar =   1.025 x £438.11   =   £449.06 per tonne.

These calculations are repeated for each different bar diameter. The results are shown in Table 5.49.

Table 5.48:   Weights of rebar for Pepperbox Farm Bridge

Bar diameter (mm)	Weight (tonnes)	Percentage of the total for the structure
12	1.54	(11% of 14.0 tonnes)
16	12.60	(89% of 14.0 tonnes)
20	6.60	(29% of 23.0 tonnes)
25	4.14	(18% of 23.0 tonnes
32	8.20	(36% of 23.0 tonnes)
40	3.91	(17% of 23.0 tonnes)

Table 5.49:   Labour and material rates for rebar in Pepperbox Farm Bridge

Bar diameter (mm)	Labour cost (£)	Material cost (£)
12	195.30	449.06
16	151.20	415.24
20	126.00	415.24
25	19.70	416.43
32	107.10	427.11
40	94.50	436.38

From consideration of Tables 5.48 and 5.49 it is possible to calculate the weighted average rate per tonne for reinforcement of 16 mm diameter and less and also for reinforcement of 20 mm diameter and over.

The total weighted average rate for 16 mm reinforcement and under = weighted average labour rate + weighted average material rate.

Weighted average labour rate
=   (0.11 x 195.30) + (0.89 x 151.20)
=   21.48 + 134.57
=   £156.05/tonne.

Weighted average material rate
=   (0.11 x 449.06) + (0.89 x 415.24)
=   49.39 + 369.56
=   £418.95/tonne.

Therefore total rate per tonne = £156.05 + £418.96 = £575.01/tonne.

A similar rate may then be calculated for the bill item relating to mild steel reinforcement of diameter greater than or equal to 20 mm. All plant requirements for this calculation were included in the preliminary item.

CONCRETE

The estimate must include the cost of providing, transporting and placing the concrete in the structures on the site. The volumes of concrete within the contract as described in the bill will be aggregated to find the total volume required of each type of concrete mix, for example:

Class of concrete	Volume in each class
Grade C15	336 m³
Grade C20 (300 kg/m³)	171 m³
Grade C25 (360 kg/m³)	2823 m³
Grade C30 (360 kg/m³)	5044 m³

Total concrete volume = 336 + 171 + 2823 + 5044 = 8374 m³

Concrete may be supplied by a local ready mix plant or batched on site. Both these alternatives must be costed.

Ready mixed concrete

Local ready mix plants must be contacted and quotations obtained for the delivery of each type of concrete mix. Each supplier will require details of:

- the total volume of concrete;
- the delivery programme for the job;
- the maximum size of concrete pour; and
- the average size of concrete pour.

When each quotation is received it must be compared with the others and with the cost of site mixing. Any special conditions relating to the quotations must be carefully reviewed.

Site mixed concrete

It may be necessary to purchase a new site batching and mixing plant. The cost of the batcher per week while on the contract may be estimated as follows:

New automatic batching plant £ 120,000

Finance cost say 20%   £   24,000

Cost of spares estimated at 10% per annum of total capital investment, including spares:

:	£ 12000	Year 1
:	£ 13200	Year 2
:	£ 14520	Year 3
:	£ 15972	Year 4
:	£ 17569	Year 5

Total cost of spares   =   £12000 + £13200 + £14520 + £15972 + £17569

                       =   £73261

Total cost of batcher =   £120000 + £24000 + £73261

                       =   £217261

Total cost written off over 5 years	=	£ 43452.20 per annum
Assuming 52 weeks/year	=	£ 835.62 per week
Therefore cost of new batcher	=	£ 835.62 per week

This cost must be compared with the cost of hiring a suitable batcher from the company plant division or other suitable supplier. If the estimator has been quoted £450 per week for a similar batching plant it would clearly appear that this is the better alternative.

The total cost of operating a site batching plant includes:

- cost of site preparation and reinstatement;
- transport to site;
- erection of batcher;
- total hire cost for the contract;
- dismantling cost;
- transportation from site;
- cost of mixer operator;
- cost of power to batching plant;   and
- cost of transporting the concrete.

Site preparation

Assume 400 m² of site must be cleared and a suitable standing laid. This will take a Drott and two labourers 10 hours of work.

Cost of labourer   =   £5.10 per hour

Cost of Drott   =   £14.37 "   "

Total cost site preparation   = (2 x £5.10 x 10) + (£14.37 x 10)

                = £102 + £143.70 =   £245.70

Assume 80 m³ of dry lean concrete is laid in a 200 mm thick layer as a hard standing for the batching plant.

Cost of the material   =   £18.00/m³

Material cost   = 80 x £18.00 = £1440.00

Assume the laying of the material will require a Drott dozer and Bomag roller with two labourers for two days working 10 hours per day.

Cost of Drott =   10 hours/day x £14.37 x 2 days   = £287.40

Roller   =   10 hours/day x £5.42 x 2 days   = £108.40

Labourer   =   2 x 10 hrs/day x £5.10 x 2 days = £204.00

Allowing £400 to cover the cost of reinstating the area, the total cost to prepare and reinstate the site of batching plant

= £245.70+ £1440.00 + £287.40 + £108.40 + £204.00 + £400.00

= £2685.50.

Transport of batcher to site

Quotation received for the transportation of the batching plant to site £750.00.

Erection of batching plant

Assume six-man gang plus a 25-tonne crane are required for three days.

Cost of one man per day = 10 hours @ £5.10 = £51.00

Cost of 25 tonne crane   = £193.16 per day

Total erection cost   = £51.00 x (6 x 3) = (3 x £193.16)

                         = £1497.48

Hire cost

Hire cost of the batching plant in operation   =   £450.00 per week.

Total cost for 70 weeks on site = 70 x £450.00   =   £31500.00

Dismantling cost

Assume a six-man gang and a 25-tonne crane are required for 2 days.

Total cost = 6 x 2 x £51.00 + (2 x £193.16) = £998.32

Transport of batcher from site

Quotation received for transportation of batching plant from site   £750.00

Cost of mixer operator

Cost of mixer operator = £7.27/hour = 50 x £7.27 = £363.50/week

Total cost for 70 weeks = £363.50 x 70 = £25445.00

Power consumption

Assume a power cost of £120.00 per week.

Total cost for 70 weeks = £120.00 x 70 = £8400.00

Cost of transporting the concrete

The cost of transporting concrete from the batching plant to the part of the site where it must be placed will depend on the nature of the work, the site and the concreting programme. Several different alternatives may be costed out.

In this example, assume that it is decided to use two concrete trucks for site transport.

Concrete trucks @ £9.54/hour   =   £9.54 x 50 =   £477.00/week

Truck driver   @ £6.75/hour   =   £6.75 x 50 =   £337.50/week

Fuel consumption - assume an average 16 litres/hour @ £0.30/litre.

Fuel cost per week =   16 x 50 x 0.30 = £240.00/week

Total cost = £477.00 + £337.50 + £240.00 =   £1054.50

Cost of two trucks for 70 weeks =   2 x £1054.50 x 70 = £147630.00

Total cost of mixing and transporting

From the above calculations it may be seen that the total cost to mix and transport concrete is as follows:

Site preparation and reinstatement	= £	2685.50
Transportation of batcher to site	= £	750.00
Erection of batcher	= £	1497.48
Total hire cost	= £	31500.00
Dismantling cost	= £	998.32
Transport of batcher from site	= £	750.00
Cost of mixer operator	= £	25445.00
Cost of power	= £	8400.00
Cost of transporting concrete	=	£ 147630.00
Total cost	=	£ 219656.30

This represents a cost of £219656.30/8374

= £26.23 per m^3 of concrete

The cost of each concrete mix

For each different concrete mix it is now possible to calculate the total cost of providing concrete to the part of the site where it must be placed. The cost of the materials in each mix may be calculated on a unit rate basis. To this must be added the batching and transportation cost, for example:

CONCRETE GRADE C30

Cost of 1 m^3 of concrete:

Cement 360 kg	@ £ 72.82/tonne	= £ 26.22
Sharp sand 595 kg	@ £ 9.20/tonne	= £ 5.47
20 mm shingle 1175 kg	@ £ 12.20/tonne	= £ 14.33
Cost of mixing and transportation per m^3		= £ 26.23

Total cost of mix = £26.22 + £5.47 + £14.33 + £26.23 = £ 72.25/m^3

It is this cost per cubic metre that must be compared with the ready mix plant quotations and a decision must be reached as to whether site batched or ready mixed concrete is to be used.

The cost of placing concrete

Irrespective of the method of concrete production the estimator must calculate the cost of placing the concrete in each part of each structure within the contract. Different labour and plant configurations for this purpose will be costed based upon an assumed daily rate of placing concrete. Whether the cost of the plant for placing concrete is entered against the appropriate bill items or included within the general section of the bill depends on the approach of the company. Where plant is specifically required for concreting only and the estimator is confident that the quantities listed in the bill are accurate then the cost of the plant may be included within the bill item rate. Where concreting plant is being utilised for other operations or the estimator doubts the reliability of the quantities in the bill then the cost should be included in the general section of the bill. Two examples of calculating the cost of placing concrete are now given.

Concrete in foundations

Assume one crane and two skips are to be used. The maximum quantity of concrete placed per day = 70 m^3

Labour cost

Assume gang consists of a concrete ganger and five labourers working an average of 10 hours per day.

Concrete ganger cost   = £ 40.60/day

5 labourers cost   =   £255.00/day

Total labour cost per day   =   £255.00 + £40.60 = £295.60

                                        at an average rate of 70 m³/day.

Labour cost   =   £295.60/70 = £4.22/m³

Plant cost

Cost of cranage and skips   =   £58/day

Cost of crane driver   =   £72/day

Vibrators and other small plant   =   £18/day

        Total plant cost per day   =   £ 148.00

Plant cost = £148.00/70   m³/day   =   £   2.11/m³

Total cost

Total labour and plant cost for placing concrete for foundations

    =   £4.22   +   £2.11   =   £6.33/m³

Placing concrete to bridge deck of Pepperbox Farm Bridge

Total pour size is 440 m³; assume that this is placed in one 12 hour shift
with a ganger and twelve men. Two concrete pumps are used and brought
on to site for the purpose of the pour.

Labour cost

Ganger 12 hour shift @ £6.69 =   12 x £6.69   = £   80.28

12 man gang for 12 hour shift =   12 x 12 x 5.1   = £ 734.40

Total labour cost per day   =   £80.28 + £734.4 = £ 814.68

(An allowance for overtime is included in the preliminaries)

Labour cost per cubic metre of concrete   =   £814.68/440 =   £1.85/m³

Plant cost

Cost of concrete pump per day   =   £350.00

Transport costs to site and return   =   £130.00

Cost of two pumps on site for one day   = 2 x £130.00 + 2 x £350.00

$\qquad\qquad\qquad\qquad\qquad\qquad\qquad\quad$ = £960.00

Total plant cost   = £960/440

$\qquad\qquad\qquad\qquad\qquad\qquad\qquad\quad$ = £2.18/m³

Total cost

Total labour and plant cost for placing concrete to the bridge deck

$\qquad$ = £1.85 + £2.18 = £4.03/m³

Example of calculation

From the assembled data the estimator may calculate the cost of providing and placing concrete for any bill item relating to any concrete on the structure. For example, assume the estimator is required to price the bill item relating to the provision and placing of concrete blinding: in situ concrete Grade C15 in blinding 75 mm or less in thickness. Concrete is to be provided as ready mixed concrete at a cost of £37.02/m³ and placed by crane and skip with a gang of six men.

$\qquad$ Labour cost = £4.22/m³

$\qquad$ Plant cost = £2.11/m³

Assume 10% wastage on the concrete, then total cost of item per cubic metre

= (1.1 x 37.02) + £4.22 + £2.11 = £47.05/m³

SITE OVERHEADS OR ON-COSTS

The calculation of the sum of money required to cover site overheads or on-costs is based upon detailed check lists containing references to all items of materials, plant and services necessary to operate a construction project.

$\qquad$ Allowance has to be made for the following site on-costs:

* supervision
* clerical staff
* site offices and compound facilities
* services
* site transport
* sundries
* general plant and major plant items.

Sample check lists for this purpose are contained in Tables 5.50 to 5.55. Examples are given below for the calculations of on-costs relating to:

* site management and supervision;   and
* the erection, maintenance and dismantling of site offices.

Check lists for on-costs

Table 5.50:   Check list for supervision

*   Project Manager
*   Agent
*   Sub-agent
*   Section Engineer
*   Assistant Engineer
*   Junior/Setting out Engineer
*   Measurement Engineer
*   Quantity Surveyor
*   Assistant Quantity Surveyor
*   Senior Laboratory Engineer
*   Laboratory Technicians
*   Works Manager
*   General Foreman
*   Section Foreman
*   Foreman - Carpenter
*   Foreman - Scaffolder
*   Foreman - Steel Fixer
*   Foreman - Bricklayer

Table 5.51:   Check list for clerical staff and other general employees

*   Office Manager
*   Cashier
*   Wages Clerk
*   Timekeeper
*   Storeman
*   Goods-received   Clerk

*   Purchasing Clerk
*   Materials Checker
*   Plant Checker
*   Cost Clerk
*   Bonus Clerk
*   Typist/telephonist
*   Industrial Relations Officer
*   Safety Officer
*   Security Staff
*   Nightwatchman
*   Chainman
*   Office Orderly

**Table 5.52:   Check list for site offices
and compound facilities**

*   Site offices
*   Section offices
*   Stores
*   Fitters' shop
*   Canteen
*   Portable offices
*   Portable shelters
*   Latrines
*   Toilet block
*   Drying-out buildings

Allowance must be made for:

*   The purchase/rental of the facilities
*   Transport to site
*   Erection and dismantling of facilities
*   Cleaning of the offices
*   All office furniture and equipment
*   Rates due
*   Construction of the compound
*   Fencing to the compounds

Table 5.53:   Check list for services

---

* Telephone installation and charges
* Electricity connection and charges
* Water installation and charges
* Drainage

Table 5.54:   Check list for site transport

---

* Staff Land Rovers
* Staff vans
* Fitters' Land Rovers
* Welding truck
* Water bowsers
* Fuel bowser
* Site lorries
* Site dumpers
* Tractor and trailer
* Staff cars (not personal)
* Resident Engineer's transport

Included for each of the above items should be:
* The cost of purchase
* Maintenance
* Insurance and tax
* Fuel and oil
* Drivers (where applicable)

Table 5.55:   Check list for sundries

---

* Surveying equipment - levels, theodolites, etc.
* Stationery, postage and telephone charges
* The provision of mobile canteen facilities
* First aid and welfare charges
* Industrial relations charges

Example calculation for site management and supervision on-costs

Fig. 5.20 shows a set of calculations for the site management and supervision required on the project. This is prepared in the following manner.

The estimator will refer to the site plan, the contract programme for the durations of activities and the direct cost estimate to ascertain:

* the nature of the site (dispersed or compact);
* the number of operatives to be controlled;   and
* the number of sub-contractors to be co-ordinated.

A formal organisation structure should be prepared for the contract and a bar chart of staff requirements prepared to ensure continuity of supervision over the duration of the contract. Fig. 5.21 is an example of such a bar chart for the contract. All technical staff, foremen and clerical staff necessary for the running of the site should be included.

In drawing up the bar chart for site supervision it is assumed that:

* the agent will be responsible to a project manager resident at head office;
* initial setting out of the main carriageway and structures will be performed by a sub-contractor survey team;
* the project will be split into two main functional sections;
* all site testing of materials will be performed by an outside company, therefore no site laboratory will be required.

Example calculation for construction of the site offices

Preparation of the area

Levelling of ground and preparation of the site will be required.

Total area = 2300 m^2

Allow a total of one week for this operation

Plant required   -   CAT 977   @   £ 26.94/hour

-   Roller   @   £ 6.28/hour

Labour required -   Banksman   @   £ 5.10/hour

-   Driver   @   £ 5.10/hour

Assume a 50-hour week

Total plant cost = (50 x £26.94) + (50 x £6.28) = £1347.00 + £314.00

= £1661.00

	No.	Weeks	Cost per week	Total Cost
1. Site Agent	1	90	460	41 400
2. Sub Agent	1	90	400	36 000
3. Section Engineer (Carriageway)	1	86	340	29 240
4. Section Engineer (Structures)	1	58	340	19 720
5. Setting Out Engineer	1	90	280	25 200
6. Setting Out Engineer	1	90	280	25 200
7. Assistant Engineer (Carriageway)	1	58	280	16 240
8. Assistant Engineer (Structures)	1	62	310	19 220
9. Assistant Quantity Surveyor	1	90	310	27 900
10. Assistant Quantity Surveyor	1	90	310	27 900
11. General Foreman	1	90	450	40 500
12. Section Foreman (Carriageway)	1	90	400	36 000
13. Section Foreman (Structures)	1	56	400	22 400
14. Foreman – Carpenter	1	56	340	19 040
15. Foreman – Steelfixer	1	56	340	19 040
	CARRIED TO SUMMARY			£405 000

Fig. 5.20   Contract supervision costs

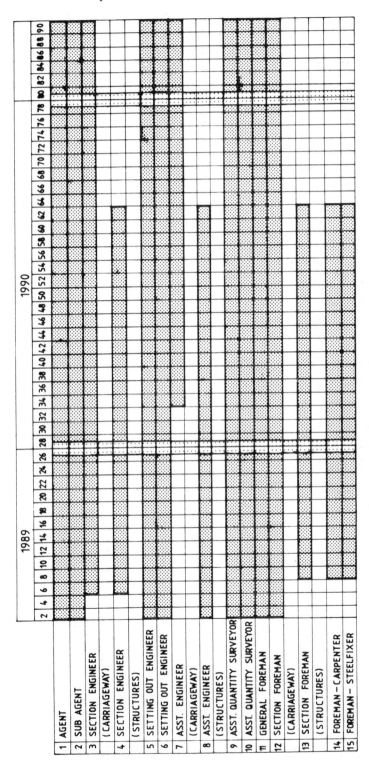

Fig. 5.21 A bar chart of supervision requirements for the contract

Total labour cost = (50 x £5.10) + (50 x £5.10) = £255.00 + £255.00

                      = £510.00

The total cost of preparation = £1661.00 + £510.00 = £2171.00

Sub-base

Sub-base to the car park and under part of the offices:

Total area	=	1000 m²
Depth of sub-base	=	0.200 m
Cost of sub-base	=	£34.00/m³ (labour and plant inclusive)
	=	1000 x 0.2 x £34.0 = £6800.00

Drainage

Installation of two septic tanks, three manholes and associated pipework:

(Total length of pipework = 200 m)

Cost of septic tanks = £850.00 each

Installation will require 12 hours of Hymac excavator @ £21.40/hour for excavation and backfill purposes, and 42 hours of labour.

The total cost of the septic tanks = cost of material + labour cost + plant cost:

Cost of tanks	=	2	x £ 850.00	=	£ 1700.00
Labour cost	=	42	x £ 5.10	=	£ 214.20
Plant cost	=	12	x £ 21.40	=	£ 256.80
Total cost	=	£1700 + £214.20 + £256.80 = £2171.00			

Cost of associated pipework = £10.12 per metre run (calculated in a similar way to the example in the drainage section)

Cost of three manholes = £781.43 each (calculated in a similar way to the example in the drainage section)

Therefore the total cost of the drainage is as follows:

Septic tanks + pipework + manholes

    = £2171.00 + (200 x £10.12) + (3 x £781.43)

    = £2171.00 + £2024.00 + £2344.29

    = £6539.29

The cost of the offices

The capital cost of the offices = £50000.00

Assume 25% is written off against the contract, i.e. 0.25 x £50000.00 = £12500.00.

The offices will be set up on concrete pads with brickwork. This will involve 54 hours of labour @ £5.10/hour and £500 worth of materials.

$$
\begin{aligned}
\text{Total cost of the offices} \quad &= \quad £12500 + (54 \times £5.10) + £500 \\
&= \quad £12500 + £275.40 + £500 \\
&= \quad £13275.40
\end{aligned}
$$

Total cost

The total cost for the construction of the offices = preparation cost + sub-base cost + drainage cost + the cost of the offices. = £2171.00 + £6800.00 + £6539.29 + £13275.40 = £28785.69

GENERAL PLANT

Included in the total for the site overheads should be an allowance for all general plant items or major items of plant, the cost of which has not been apportioned to individual bill items. The quantities stated against each item in the bill may vary with those actually required for the completion of the work. If the quantity is reduced from that stated in the bill, the contractor may not recover the money for a major item of plant if the cost has been included within the item rate. A smaller item quantity will mean a reduction in payment for the operation which may not be proportional to the cost saving obtained. Therefore, it is safer to include major plant items in the on-costs.

Typical plant items that may be included are listed in Tables 5.56 to 5.62, which are categorised under the following headings:

* Plant, tools and equipment
* Plant haulage
* Cranage
* Scaffolding
* Provision for safety
* Protection against noise
* Temporary works

Having calculated the direct cost of performing the works, should the contract be awarded the estimator must assemble the direct cost estimate in a manner which may be extended and discussed at the tender adjudication meeting. This requires the preparation and presentation of the reports described in Chapter 2 and discussed in detail in Chapter 6.

Table 5.56:   Plant, tools and equipment

* Batching plant
* Running gear
* Generators
* Crushing plant
* Plant for precast yard
* Small mixers
* Conveyors
* Compressors
* Barbending plant
* Traffic lights
* Traffic light control units
* Diesel tanks
* Petrol tanks
* Weighbridge
* Consumables
* Protective clothing

Table 5.57:   Plant haulage

* Excavators
* Cranage
* Tractors and scrapers
* Offices/stores
* Batching plants

Table 5.58:   Cranage

* Tower cranes
* Mobile cranes
* Derrick cranes
* Kentledge, etc.
* Erection of cranes
* Dismantling of cranes

Table 5.59:   Scaffolding

* External
* Internal
* Soffits
* Towers
* Hoists
* Labour

Table 5.60:   Provision for safety

* Handrails to staircases
* Rails to roof edges
* Temporary covers
* Additional lighting
* Netting
* Protection to hoists
* Safety boats (river work only)

Table 5.61:   Protection against noise

* Noise barriers/screens
* Proprietory piling equipment
* Soundproofing to compressors

Table 5.62:   Temporary works

---

* Access road
* Tip areas
* Pumping equipment
* Dewatering equipment
* River/stream diversions
* Cofferdams
* Bailey bridges
* Site lighting
* Weather protection
* Temporary fencing
* Publicity
* Advertising

---

6 Calculations and Decisions in Tendering

ESTIMATING AND TENDERING

Estimating consists of calculating the probable cost to the contractor of efficiently carrying out the construction work if awarded the contract to construct the project. Tendering consists of establishing the final price and terms for the contract that will be submitted to the promoter or his representative. This involves an assessment of the likely margin of error in the estimate together with the risk and possible financial effects of undertaking the project.

A tender meeting should ideally take place a few days before the tender is due to be submitted. The purpose of this meeting is to review the estimate and finalise the contract price in order that the final tender rates for the bill items may be calculated and the tender documents prepared for submission. Often commercial pressures result in the tender meeting taking place the day before tender submission. This produces additional pressures on the estimator and senior management.

INFORMATION AND REPORTS FOR THE TENDER MEETING

As described in Chapter 2, the estimator prepares reports for the tender meeting that will enable senior management to understand the project fully and assess the risks involved in order to determine an appropriate tender price. By this stage the estimator will have built up a detailed knowledge of the project and this is conveyed to the senior management through a series of reports and verbal explanations. Reports will be presented to senior management under the two general headings of:

* the construction of the project; and
* contractual and commercial details.

The construction of the project

The information that will be given will include:
* a description of the works;
* the report of the estimator's site visit;
* the method statement for the construction of the works;
* the programme for construction;
* major assumptions made in the preparation of the estimate; and
* details of any unresolved technical problems.

Contractual and commercial details

Information will be required by the estimator on the following contractual aspects:
* the client and his advisors involved in the project;
* the conditions of contract;
* the terms and quotations relating to materials suppliers and sub-contractors;
* relevant unresolved contractual problems;
* the potential profitability of the contract;
* any need for explanatory letters or qualifications to the tender; and
* the market and industrial conditions relating to the project including details of other tenderers.

Typical reports

Typical reports supplied by the estimator for the tender adjudication meeting include:
* an outline report of the project, a typical example was given in Chapter 5;
* a detailed site report, a typical example was given in Chapter 5;

* a bar chart programme and method statement, a typical example was given in Chapter 5;
* a list of labour totals with the total number of hours given in each of the main categories of labour, see Table 6.1;
* a list of plant requirements, see Fig. 6.1;
* labour and plant resource requirements, typical examples are given in Figs 6.2 and 6.3;
* a list of sub-contractors and the total value of their work, see Table 6.2;
* a copy of the bill of quantities giving the direct cost rates, a typical page is shown in Fig. 3.4;
* a report summarising the costs in each trade within the contract, see Table 6.3;
* a summary of the cost of each section of the bill, see Table 6.4; and
* details of the cash flow for the project, see Figs 6.4, 6.5 & 6.6.

The figures given within these tables are not those from one particular contract and should not be interpreted as such. They have been constructed with artificial values to show the typical format of the reports.

The estimator will supply a list of labour resources and the total number of hours required for each resource. An example is given in Table 6.1.

Table 6.1   Labour requirements for the contract

Labour resource	Hours
Labourer	132000
Plant operator	46200
Steel fixer	29100
Carpenter	25462
Pipelayer	7500

Table 6.2 shows details of sub-contractor requirements for the contract.

ITEM		QUANTITY			FUEL		COST RATES			COST TO CONTRACT			
Description	H or C	No	Time in Days	Total Plant Days	Litres per Day	Total Fuel	Machine per Day	Cost of Fuel per Litre	Operator per day	Machine	Fuel	OP	Total
22RB	C	6	Varies	1540	32	49,280	134.64	0.34	58.16	207,345	16,755	89,566	313,666
JCB 3C	H	2	175	350	26	9,100	159.20	0.34	INCL	55,720	3,094	INCL	58,814
HYMAC	H	2	380	760	25	19,000	171.20	0.34	INCL	130,112	6,460	INCL	136,572
CAT D6	H	1	180	180	40	7,200	257.76	0.34	58.16	46,396	2,448	10,468	59,312
DROTT	H	1	180	180	32	5,760	114.96	0.34	58.16	20,692	1,958	10,468	33,118

Note: The plant item may be a company plant item or hired from a plant hire company.
This is shown as   C = Company   H = Hire plant

Fig. 6.1 A list of plant requirements

Table 6.2   Sub-contractor requirements

Operation	Company	Sum £	Disc %	Net sum £
Fencing	Thorpe & Son Ltd	137600	5	130720
Street lighting	Siddons Smith	133857	-	133857
Surfacing to carriageways	Tilmac Roadstone Ltd	2838000	-	2838000
Handrailing	Varley and Gulliver Ltd	31419	3	30476
Landscaping	Bond Ltd	85797	5	81507

All the above sub-contractors were selected by the main contractor. A separate list would show nominated sub-contractors and suppliers.

The contract direct cost total may be summarised by the various trades as shown in Table 6.3 below.

Table 6.3   Trade summary for the contract

Trade	Labour	Plant	Mats	S/C	Total
Demolition/ Site clearance	2040	4250	-	-	6290
Earthworks	139581	620850	21260	-	781691
Drainage	180540	215560	1373589	-	1769689
In-situ concrete	206476	88916	118155	-	413547
Precast concrete	-	7344	214200	-	221544
Formwork	173741	21234	116111	-	311086
Reinforcement	197880	47941	155551	-	401372
Landscaping	-	-	-	81507	81507
Roads & pavings	105581	217773	138029	145350	606733
Others	166778	105576	325213	332865	930432

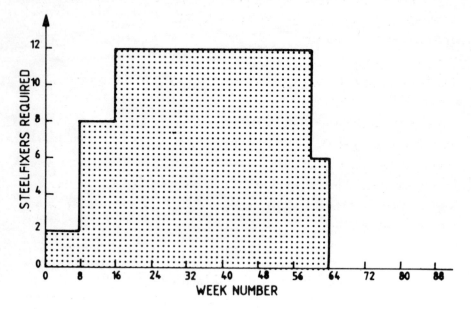

Fig. 6.2 Example labour requirements

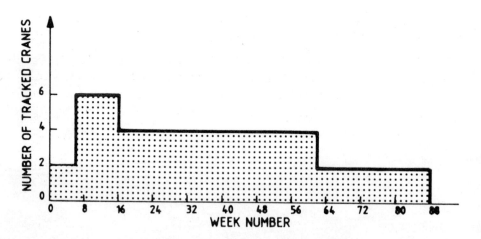

Fig. 6.3 Example plant requirements

Alternatively the summary may be prepared on a section by section basis as shown for another contract in Table 6.4.

THE CASH FLOW ANALYSIS FOR THE CONSTRUCTION PROJECT

The cost of financing a construction project should be included in the mark-up additions for head office overheads. This can be calculated from the project cash flow. An example of the cash flow analysis for a typical construction project is as follows. Because of the time allowed the estimator for this task, often it is only possible for an approximate analysis of the cash flow for the project to be produced.

Table 6.4   Direct cost summary section totals

Bill	Sect	Labour	Plant	Materials	S/C	Prov/PC	Total
1	1	1000	980	712	-	42000	44692
	2	4000	29490	6712	1200	-	41402
	3	16819	18043	12153	16741	-	63756
Total		21819	48513	19577	17941	42000	149850
2	4	21012	21492	7600	4398	5000	59502
	5	7153	36075	15071	1000	-	59299
	6	6131	16123	10102	26000	12000	70356
	7	12010	8941	19103	4100	-	44154
Total		46306	82631	51876	35498	17000	233311
3	8	9812	29713	17400	18267	-	75192
	9	4076	32104	8132	77824	-	122163
	10	5100	16000	7140	49822	-	78062
Total		18988	77817	32672	145913	-	275417

To calculate the cash flow the following data are needed:

- a graph of contract value against time;
- a graph of contract cost against time;
- the measurement and certification interval;
- the payment delay between certification and the contractor receiving the cash;
- the retention conditions and repayment arrangements;
- the project costs broken down into labour, plant, materials and sub-contractor categories;   and
- the terms of paying the resource costs and the delay between incurring a cost liability and making payments.

Value versus time

A graph or set of figures of the contract value against time is required to represent the amount of work being done in each time period in terms of the value earned. Within the estimate and tendering calculations the total tender sum is known and can be broken down into various sections such as a bill section or a particular trade. It cannot, however, be broken down into time periods without additional work. The correct procedure is to use the pre-tender programme or bar chart and to value each activity or bar by the bill rates (including the mark-up for overheads and profits). This is to consider each bar to be a collection of bill items. Fig. 6.4 gives an example of a bar chart with each bar value. Fig. 6.5 shows the derived value versus time graph from which the contract income can be calculated. The graph in Fig. 6.5 is similar to the set of figures in Row 1 of Fig. 6.6.

At the time the estimator carries out these calculations the only bill rates available to him are likely to be the direct cost rates. The decisions on additions for risk, overheads and profits may not have been taken. Therefore, estimators are sometimes required to prepare this for a range of assumed mark-up rates to calculate the consequences of different mark-up decisions on the cash flow.

The following figures represent a considerable amount of calculation and if it is repeated several times for different mark-up rates the effort required can be prohibitive. The use of computer based 'spreadsheets' allows these calculations to be performed much more easily than could be achieved

	WEEK NUMBER																			
	1	2	3	4	5	6	7	8	9	10	11	12	13	14	15	16	17	18	19	20
EXCAVATE	5	5																		
CONC. FOUNDNS		2.5	10																	
G.F. SLAB			2.5	5																
COLS G TO F1				10																
FLOOR 1					5	10														
COLS 1 TO F2							10													
FLOOR 2								7.5	10											
COLS 2—ROOF										10										
ROOF STRUCTRE											5	5								
CLADDING											17.5	5	2.5	2.5						
SERVICES													10	2.5	2.5					
FINISHES															2.5	2.5				
VALUE £(0000's) PER MONTH	5	7.5	12.5	5	10	5	10	10	7.5	10	27.5	10	17.5	5	5	2.5				
CUMULATIVE VALUE £(0000's)	5	12.5	25	30	40	45	55	65	72.5	82.5	110	120	137.5	142.5	147.5	150				

Fig. 6.4   Bar chart programme with contract values

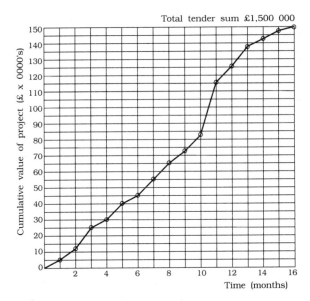

Total tender sum £1,500 000

Fig. 6.5   Value vs. time graph from bar chart

		1	2	3	4	5	6	7	8	9	10	11	12	13	14	15	16	17	22
1	CUMULATIVE VALUE	5.50	12.50	25.0	30.0	40.0	45.0	55.0	65.0	72.5	82.5	110.0	120.0	137.50	142.5	148.0	150.0		
2	CUMULATIVE VALUE RETENTION	4.75	11.87	23.75	28.5	38.0	42.75	52.25	61.75	68.87	78.37	104.5	114.0	130.62	135.37	140.60	142.50		
3	CUMULATIVE PAYMENT REC'D FROM CERT		4.75	11.87	23.75	28.5	38.0	42.75	52.25	61.75	68.87	78.37	104.5	114.0	130.62	135.37	140.60	142.50	142.50
4	CUMULATIVE RETENTION PAYMENT																	3.75	7.50
5	CUMULATIVE COSTS	4.35	10.87	21.75	26.10	34.80	39.15	47.85	56.55	63.07	71.77	95.70	104.40	119.62	123.97	128.32	130.50		
6	CUMULATIVE LABOUR COSTS	0.87	2.175	4.35	5.22	6.96	7.83	9.57	11.31	12.61	14.35	19.14	20.88	23.92	24.79	25.66	26.10	26.10	
7	CUMULATIVE LABOUR PAYMENT	0.87	2.175	4.35	5.22	6.96	7.83	9.57	11.31	12.61	14.35	19.14	20.88	23.92	24.79	25.66	26.10		
8	CUMULATIVE MATERIAL COST	0.87	2.175	4.35	5.22	6.96	7.83	9.57	11.31	12.61	14.35	19.14	20.88	23.92	24.79	25.66	26.10		
9	CUMULATIVE MATERIALS PAYMENTS		0.87	2.175	4.35	5.22	6.96	7.83	9.57	11.31	12.61	14.35	19.14	20.88	23.92	24.79	25.66	26.10	
10	CUMULATIVE PLANT COSTS	1.74	4.35	8.70	10.44	13.92	15.66	19.14	22.62	25.2	28.71	38.28	41.76	47.85	49.59	51.33	52.20		
11	CUMULATIVE PLANT PAYMENTS		1.74	4.35	8.70	10.44	13.92	15.66	19.14	22.62	25.2	28.71	38.28	41.76	47.85	49.59	51.33	52.2	
12	CUMULATIVE S/C COSTS	0.87	2.175	4.35	5.22	6.96	7.83	9.57	11.31	12.61	14.35	19.14	20.88	23.92	24.79	25.66	26.10		
13	CUMULATIVE S/C PAYMENTS		0.87	2.175	4.35	5.22	6.96	7.85	9.57	11.31	12.61	14.35	19.14	20.88	23.92	24.79	25.66	26.10	
14	CUMULATIVE CASH OUT	0.87	5.65	13.05	22.62	27.84	35.67	40.91	49.59	57.85	64.77	76.55	97.44	107.44	120.48	124.83	128.75	130.5	130.5
15	CUMULATIVE CASH FLOW	-0.87	-0.9	-1.18	+1.13	+0.66	+2.33	+1.84	+2.66	+3.90	+4.10	+1.82	+7.06	+6.56	+10.14	+10.54	+11.85	+15.75	+19.5

Fig. 6.6 Calculation of a contract cash flow

manually. Alternatively short cut calculations are used. The most common short cut is to assume the shape of the value versus time graph. Guidance on the shape of these curves can be obtained by examining previous contracts and reducing the value versus time graph to percentage scales;   i.e. by assuming that at 100% of time, 100% of value has been earned. This will show that there is some consistency in the shape of these graphs. By knowing the tender sum (cost estimate + mark-ups) and the overall duration, the assumed percentage graph can be scaled up to absolute values of money and time.

Cost versus time

To produce a similar graph for cost versus time simply 'cost' each bar in the pre-tender bar chart by using the direct cost rates. This will produce a graph similar to the value versus time but representing costs. The figures in Row 5 of Fig. 6.6 are an example of costs versus time.

The short cuts are used to produce the value versus time graph by means of percentage curves derived from previous contracts and to adjust these to give an approximation to the cost versus time curve.

Assuming each bill item carries the same percentage additions then the cumulative cost against time is a simple proportion of the cumulative value for each month.   For example, if the contribution margin is 15%:

$$\text{Value} = \text{cost} + 15\% = 1.15 \times \text{cost}$$
$$\text{Therefore cost} = \text{value}/1.15 = 0.87 \times \text{value}$$

Costs in each category

The previous section outlined how the overall costs are apportioned over time. It is necessary to do this for each cost category. The estimate will give the total cost of each of the categories (labour, plant, materials, sub-contractors, etc.) but like the overall costs it is not available week by week. To allocate these costs to time is an enormous task. Again computer based 'spreadsheets' may be used to help the estimator achieve this task. Without computers the estimator is forced to use short cuts. The most common short cut is to break the total cost in each time period into cost categories in the same proportion as the overall contract totals.

From the direct cost summary for the contract it is possible to calculate the percentage cost in each resource category: labour, plant, materials, etc. For this project assume:

	%	£
Labour	20	261000
Plant	40	522000
Materials	20	261000
Sub-contractors	20	261000
Total	100	1305000

The costs in each time period can now be divided into these proportions.

The measurement and certification interval and the payment delay between certification and the contractor receiving cash

These details are required to convert the value versus time data into income. The following are examples of typical values:

Retention will be 5% of the amount paid to the contractor up to 3% of the tender total.   One half of the total retention money will be paid on completion of the works, and the remainder paid after the end of the period of maintenance.

The first certificate will be submitted at the end of the first month of the project, and thereafter on a monthly basis. The delay between submission and payment will be fourteen days which will be taken as one month in these calculations.

These items of data are based upon those stated in the Conditions of Contract for Works of Civil Engineering Construction (8).

Payment by the contractor of resource costs

The delay between receiving goods or services and paying for them is an important factor in the calculation of the cash flow. The delays between using labour, plant, materials and sub-contractors are determined by current trading conditions.

The following periods are included in the example:

* Labour   : Nil
* Plant   : One month
* Materials   : One month
* Sub-contractors   : One month

Calculation of the cash flow

Having assembled the above data it is possible to calculate the cash flow for the contract. This is shown in Fig. 6.6. In this figure the time period of one month is used. (A more accurate calculation could be obtained by using a time interval of one week.) The explanation of Fig. 6.6 is as follows:

Row 1 is the cumulative value for the project derived as previously described.

Row 2 is the cumulative value less retention.

Row 3 is the value less retention (Row 2) shifted by an amount equal to the payment delay between valuation and the contractor receiving his money.

Row 4 is the cumulative retention payments inserted at the time they would be received.

The cumulative costs for the total project are shown in Row 5. The proportion of costs due to labour is calculated and inserted in Row 6. This is repeated for materials in Row 8, for plant in Row 10 and for sub-contractors in Row 12.

Rows 7, 9, 11 and 13 shift the cost liabilities shown in Rows 6, 8, 10 and 12 by an amount equal to the average delay in incurring the cost liability and making the payment.

Row 14 is the cumulative cash out and is the sum of Rows 7, 9, 11 and 13.

Row 15 is the cumulative cash flow from Row 3 plus Row 4 less Row 14. Row 15 represents the difference between cash in and cash out throughout the contract. Row 15 shows the maximum negative cash flow, i.e. the lowest amount of cash required to fund the contract. It also shows the time at which the contract becomes self funding. Using these data or the curves plotted on Fig. 6.7 the interest charges on the negative cash flows can be calculated.

The discrepancy between costs and revenue received must be met by the contractor from capital either supplied from the company's cash reserve or borrowed. If the cash is borrowed, interest will be charged to the company; if cash from the cash reserve is used, the project should be charged for the loss of interest earning capability. The interest payable may be obtained by calculating the shaded area between the two curves in Fig. 6.7.

Area between the curves $=$ 3.4 cm²

(Where 1 cm² = 100000 £ months)

Assume interest rate $=$ 19% per annum

Interest payable $=$ $\dfrac{3.4 \times 100000 \text{ £ months}}{12 \text{ months}} \times 19\%$

$=$ £5383

An allowance in the overheads to cover this amount may then be made.

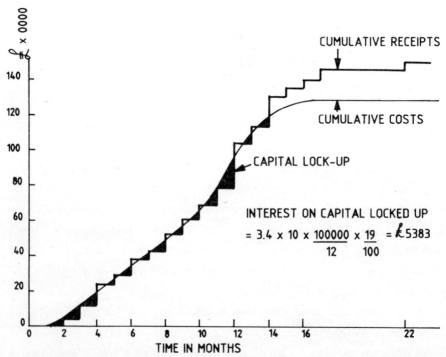

Fig. 6.7   **Cash in and cash out graphs with finance area shown**

Factors which affect contract cash flow

The factors which affect a contract's cash flow are:
* the mark-up;
* retentions;
* claims;
* rate loading (front or back);
* over or under-measurement;
* delays in receiving payment from client; and
* delays in paying for goods or services.

The effect and importance of each of these are fully described in Chapter 14 of 'Modern Construction Management' by Harris and McCaffer (1).

TENDER PANEL'S DECISIONS

The tender panel has to come to decisions with regard to:
* adjustments to the direct cost estimate;
* additions for head office overheads, profit and risk;
* adjustment to allow for inflation and rate loading; and
* writing up the bill.

Adjustments to the direct cost estimate

The direct cost estimate for the project is assessed at a tender meeting by representatives of the senior management and the estimating team. It is the responsibility of this panel to satisfy themselves that the estimate is adequate and represents the most likely cost to the company of completing the work if awarded the contract. The estimator responsible for the preparation of the estimate will supply a copy of the bill of quantities with direct cost rates together with a number of different reports as shown in Chapters 5 and 6. The estimator will be questioned as to his assumptions and decisions with respect to his selection of the:
* plant and labour for the project;
* plant and labour outputs; and
* site overhead requirements.

The construction programme for the project will be examined in detail together with the method statements to determine any omissions that must be allowed for or any savings that may be made in labour or plant requirements.   Although it may be too late in the tender period to make detailed revisions to the direct cost rates it is important that the best method of working on the project is determined and represented in the tender price.

It is usual for a list of lump sum additions and subtractions to the direct cost estimate to be prepared in order that the total sum may be adjusted. These changes will be the result of the panel decisions as to changes required in resource levels for the project and also adjustments for materials and sub-contractor prices that have arrived too late to be used in the recalculation of bill item direct cost rates.

Additions for head office overheads

The construction of the project must contribute towards the cost of running the company's head office and other general overheads of the organisation. The company's accountant will be required to submit the latest details of this cost to the company.   In order to provide accurately for overhead costs it is necessary to know the relationship between costs and turnover.

A construction company should always be tendering within a defined tendering policy.   The information the policy should provide is:

- a turnover target divided into each market showing the proportions in which the total turnover is to be obtained;
- an overheads budget;
- gross and net profit targets;   and
- the anticipated volume of enquiries required to achieve the turnover.

The tendering policy can be represented by a tendering programme, showing the values and latest dates to achieve the volume of enquiries and turnover target.

A monthly comparison of the level of turnover with overhead costs should indicate the required contribution from future contracts.   An example is given in Table 6.5.

Table 6.5   Calculation of contract overhead requirements

Total overhead cost for last financial year = £960000

Assume an increase of 10% for current year

   = 0.10 x £960000 = £96000

Total overhead with cost to be recovered

   = £960000 + £96000 = £1056000

This represents an average overhead recovery

   = £1056000/12 = £88000 per calendar month

The estimated total turnover for the current financial year

   = £17000000

Therefore the average monthly turnover

   = £17000000/12 = £1416667

The cost of the contract is estimated at £3780000

The average monthly turnover for the contract being

assessed at £180000 per month

Therefore overhead required = (£180000/1416667) x 88000

   = £11181 per month

Total overhead recovery = overhead per month x number

of months in the contract = £11181 x 21 =   £234801

As a percentage of the direct cost for the project

   = (234801/3780000) x 100 = 6.2%

Additions for profit and risk and the calculation of the total tender sum

Profit

The profit margin added to the direct cost for a particular project varies with:

*   the volume of work in hand;
*   the orders anticipated;
*   market conditions; and
*   the risk inherent in the works.

Underlying the considerations is a requirement to make an adequate return on capital invested in the company to satisfy shareholders, service debts and to re-invest in the company after paying corporation tax. Thus on considering these issues senior management will decide the profit allowance that should be added to the contract.   This may produce a tender sum considered too high to win the contract. The company must then decide by what sum they are willing to reduce the tender sum in order to obtain the contract.

Where the contractor has little work in hand there is obviously great pressure on the management to secure a contract by submitting the lowest tender. The direct cost estimate will be reduced to as low a figure as possible and the allowance for profit kept to a minimum.

Discounts taken on materials and sub-contractors' quotations are sometimes considered as an extra source of profit and it is not unknown for a contractor in very poor markets to submit a tender with zero or negative profit allowance and rely on such discounts to produce the profit required. Thus strictly speaking the discounts should be considered as part of the tender additions.

In situations where few further orders are anticipated in this area of work or where prevailing market conditions indicate that there will be many companies seeking to secure the contract the profit margin will again be kept to a minimum.

The determination of the minimum profit required by the company is largely determined by the demands placed on the profit earned. The company has to satisfy the following demands:

(1)   corporation tax, currently at 35%;

(2)   interest on borrowed capital, this varies according to interest rates;

(3)   dividend payable to shareholders on equity capital; this is determined by the Board of Directors but must be set at a level that satisfies shareholders and maintains share value of the company;   and

(4)   re-invested profits to sustain a growing company without increasing borrowings to an unacceptable level.

From a knowledge of the requirements of these factors a minimum profit/capital ratio can be established. Thus through the capital/turnover ratio the minimum profit/turnover can be determined. The following is an illustrative example.

The following assumptions are used in respect of the calculation of minimum profit required:

Equity capital in company   =   £1,000,000

Borrowed capital in company   =   £500,000

Interest on borrowed capital   15%

Intended dividend   10% on equity capital

Intended reserve profits   30%

Corporation tax   35%

Target turnover/capital ratio   8

Minimum profit required, P   =   0.15 x 500,000 (interest on borrowed capital)

+ 0.1 x 1,000,000 (dividend)

+ 0.3P (allowance for reserved profits)

+ 0.35 (P - 0.15 x 500,000) (corporation tax on profit less interest charges)

$\therefore$ P   =   75,000 + 100,000 + 0.3P + 0.35P - 26,250

$\therefore$ P   =   148,750 + 0.65P

$\therefore$ P   =   £425,000

Check: Profit   =   425,000

Corporation tax

= 0.35(425,000-75,000) =   122,500
                            ———————
                            302,500

Reserved profit

= 0.3 x 425,000   =   127,500
                          ———————
                          175,000

Servicing debt

= 0.15 x 500,000   =   75,000
                          ———————
                          100,000

Dividend

= 0.1 x 1,000,000   100,000
                           ═══════

If the turnover/capital ratio of 8 is achieved then turnover

= 8 x 1,500,000   =   £12,000,000

$\therefore$ Minimum profit   = $\dfrac{425,000}{12,000,000}$ =   3.54%

i.e. $\dfrac{\text{Profit}}{\text{Turnover}}$ =   3.54%

$\dfrac{\text{Turnover}}{\text{Capital}}$ =   8

$$\frac{\text{Profit}}{\text{Capital}} = 28.32\%$$

If the company achieved a higher turnover of £15,000,000, the turnover/capital ratio would be 10.

$$\text{Minimum profit} = \frac{425,000}{15,000,000} = 2.83\%$$

$$\frac{\text{Profit}}{\text{Capital}} = 28.3\%$$

Risk

The adjudication panel must assess any risk inherent in the project and make allowance for this in the total tender sum. The estimator should identify any physical uncertainties within the construction of the project and provide alternative methods of construction with appropriate costings. The senior management can then assess the commercial significance of these uncertainties. Although mathematical procedures exist for examining uncertainty and risk through the introduction of probability analysis it is unusual for the panel to enter into such calculations. Risk is normally allowed for by the addition of a single sum of money equal to an agreed percentage of the direct cost and included in the profit allowance. Because of the different level of risk between company work and sub-contracted work different profit margins may be applied to the company's work and sub-contractor's work.

THE PRODUCTION OF THE TOTAL TENDER SUM

Having considered the various factors described above, the panel will summarise the various additions for:
- site overheads;
- head office overheads;
- profit;
- risk;

and produce a final tender sum.

This calculation is performed on a single tender summary sheet. The exact format of the sheet will vary from company to company. A typical example is shown in Fig. 6.8.

The site on-costs may be split into sub-categories of labour, plant and materials, etc. to highlight the relationship between the direct cost in each category and the on-costs required. Overheads, profit and risk may be added as a percentage of the direct cost total or as a single sum of money. Whatever the procedure adopted for the calculations a total tender sum will be arrived at and agreed by the panel. The figures in the bill of quantities must then be adjusted to show this sum and any rate loading of the bill items that is required.

Adjustments to the tender to allow for inflation and rate loading

In 1971, the Steering Group on Price Adjustment Formulae (10) published a report recommending a formula based on the use of monthly indices as a means of adjusting contract values for changes in the cost of the labour, plant and materials required in the construction of civil engineering projects. The formula was based on monthly price indices reflecting changing costs in the following elements of construction work within civil engineering contracts:

(i)	labour;
(ii)	plant;
(iii)	aggregate;
(iv)	cement;
(v)	bricks;
(vi)	cast iron;
(vii)	coated roadstone;
(viii)	fuel;
(ix)	imported softwood;
(x)	reinforcement; and
(xi)	structural steelwork.

The Department of the Environment became responsible for collating and publishing monthly figures for indices in the above categories. The fifth edition of the 'ICE Conditions of Contract' reprinted in January 1986 (8) adopted this method of price adjustment using these indices as a replacement to the clause relating to the variation of price (labour and materials) within the fourth edition of the 'ICE Conditions of Contract'. The use of the formula has been accepted by the civil engineering industry.

The application of the civil engineering price adjustment formulae to contracts changed the approach of contractors who had previously

THORPE CONSTRUCTION LTD

Job Number: ....   Date: ....

Contract: ....   Estimator: ....

Client: ....   Planner: ....

SUMMARY ANALYSIS SHEET

Direct Costs:	SUBTOTAL	TOTAL	% OF TOTAL CONTRACT COST
Labour			
Plant			
Materials			
Company Total			
Direct Subcontractors			
Indirect Costs:			
Salaries			
Labour			
Plant			
Materials			
Subcontractors			
Total Indirect Costs			
Total Contract Cost			

Head Office Overhead .... % of ....

Profit Margin   .... % of ....

Add Provisional Sums

Add Prime Costs

Add Insurance   .... % of ....

Add Dayworks

Add Contingency Allowance

TENDER PRICE

% Gross Margin

Sum Required in the Bill   £

Amount to Spread   £

Fig.   6.8   A tender summary sheet

concentrated on reducing their capital lock up by attempting to gain payment for all works at the earliest possible opportunity. If the rate of inflation was higher than the cost of capital, then by improving their contract cash flow contractors were depriving their companies of adequate compensation as calculated by the formula for increased costs.

The factors which affect the contract cash flow have been listed in the previous section. By considering these factors the contractor may minimise his capital lock-up by increasing specific item rates within the bill of quantities while keeping the total tender sum the same by reducing others. This action is known as 'rate loading'. Rate loading may be performed on individual bill items or complete classes of work. The intention of rate loading is to affect the cash flow for the project but not the competitiveness of the tender. An example calculation in Table 6.7 shows the method of adjustment of bill rates.

Table 6.7   A bill of quantities

Item	Description of work	Unit	Qty	Rate	Amount
A	Class F3 'formwork'	m^2	8	28.25	226.00
B	High tensile rebar 20mm dia.	tonne	28.39	453.67	12879.69
C	Excavate unsuitable material	m^3	40	11.22	448.80
D	In-situ concrete Grade C15 OPC	m^2	20	119.70	2394.00
E	Two coats tar sprayed on water-proofing	m^2	303	3.20	969.60
	TOTAL TENDER SUM				16918.09

It could be decided to increase the rate for the bill item D by 20% and adjust the other bill items accordingly to keep the total tender sum the same:

i.e.   item rate (D)   =   119.70 x   1.20   =   143.64

item amount (D)   =   2872.80

Increase in total tender sum   =   2872.80 -   2394.0

=   478.80

Adjustment

for other items = $\dfrac{\text{change in item D}}{\text{total tender sum - original item D amount}}$

$= \dfrac{478.80}{14524.09}$

$= 0.033$

Adjustment   =   1 - 0.033 = 0.967
as a multiplier

This then becomes the multiplier for all bill items rates except item D, and the re-adjusted bill becomes as shown in Table 6.8.

The scale of the calculations involved would require prodigious clerical effort if all the items in a bill of quantities were to be adjusted manually. Consequently only a number of relevant bill item rates are adjusted by the estimator following the tender adjudication meeting.

Table 6.8   The bill of quantities adjusted after rate loading

Item	Description of work	Unit	Qty	Rate	Amount
A	Class F3 'formwork'	m²	8	27.32	218.56
B	High tensile rebar 20mm dia.	tonne	28.39	438.76	12456.46
C	Excavate unsuitable material	m³	40	10.85	434.00
D	In-situ concrete Grade C15 OPC	m³	20	143.64	2872.80
E	Two coats tar sprayed on water-proofing	m²	303	3.09	936.27
	TOTAL TENDER SUM				16918.09

In this example a minor adjustment has been made to the rate for B to bring the bill total close to £16918.09, this is required because of rounding errors in the arithmetic.

Writing up the bill

In Civil Engineering contracts the promoter usually requires the construction company tendering for the works to submit with the form of tender a priced bill of quantities giving the tender rates for each bill item and the total tender cost. Therefore the estimator has to produce the gross rates for the work and enter them into a copy of the bill for submission. There will inevitably be a shortfall between the sum of money contained in the bill of quantities prior to the tender adjudication and the total tender sum as finally agreed by the panel. This discrepancy may be overcome by one of two methods:

* the increase of bill item rates by a single percentage;   or
* the introduction of a single sum addition against the adjustment item within the general summary of the bill of quantities.

The increase of bill item rates is normally only applied to the company's own work. When, in the opinion of the panel, the work may have been overmeasured and there is a possibility that the quantities of certain items may be reduced during the construction of the project, more money will be introduced into the preliminary items to ensure that the costs are recovered by the company. No addition is made to provisional sums or provisional items within the bill as the company cannot be assured of their inclusion in the project when constructed.

Usually a combination of these two methods is adopted. An example is given in Table 6.9.

When the policy for marking up the bill has been decided by the panel and a priced bill is required by the client the estimator must go through the bill pages increasing the rates of the appropriate items and making the necessary adjustments. The final bill item rates must then be entered manually into an unmarked copy of the bill and this is submitted to the promoter. An example of a bill page prior to submission is given in Fig. 6.9.

Because of the very short period between the tender adjudication meeting and the submission of the tender, it is not uncommon for estimators to include a notional percentage and mark up to the bill items at the time of calculating the direct cost rates. This allows the tender rates to be calculated in advance of the tender adjudication meeting and minimises the amount of adjustment required to the bill items prior to the submission of the tender.

Table 6.9   The determination of the method of mark-up

for the contract

Total tender sum	£ 4376106
Total monies in the bill	£ 3102615
	—————
Shortfall	£ 1273491
Less site on costs	£ 507412
	—————
	£ 766079
	—————

This total as a percentage of the total company work = (766079/3829764)

$$\text{x 100}$$

$$= 20.0\%$$

Therefore mark-up strategy = enter sum of £507412 (to cover site overheads) into the general summary section of the bill as an adjustment item. Increase all company work items by 20.0% to cover remaining shortfall.

	HEADWALLS, CULVERTS AND SEWERS TO EXISTING BYE-PASS BRIDGE				
	In-situ Concrete				
	Provision of concrete; prescribed mix for ordinary concete taken from Table 50 of C.P. 110: Part 1				
F213	Grade 10; 20 mm aggregate)	m³	8	47.03	376.24
F214	Grade 10; 40 mm aggregate)	m³	52	45.53	2367.56
F233	Grade 20; 20 mm aggregate)	m³	1	50.17	50.17
F243	Grade 25; 20 mm aggregate)	m³	41	52.64	2158.24
	Placing of Concrete				
	Mass				
F611.1	Blinding 75 mm thick)	m³	1	58.00	58.00
F611.2	Blinding 75 mm thick; to slope)	m³	7	59.00	413.00
F624	Bases, footings and ground slabs) over 300 mm thick)	m³	52	68.00	3536.00
	Reinforced				
F722	Bases, footings and ground slabs) 150 – 300 mm thick)	m³	31	70.00	2170.00
F742	Walls 150 – 300 mm thick)	m³	7	74.00	518.00
F773	Casting to metal sections (pile up)) 0.1 – 0.25 m² cross sectional area)	m³	3	76.50	229.50
	Concrete Ancillaries				
	Formwork; rough finish				
G143	Plane vertical width 0.2 – 0.4 m)	m²	15	15.85	253.60
G144	Plane vertical width 0.4 – 1.22 m)	m²	23	23.30	535.90
G145	Plane vertical width over 1.22 m)	m²	21	31.70	665.70
	Carried to Collection				£13331.91

Fig. 6.9   Sample page from a CESMM2 bill of quantities prior to submission

7 Estimating and Tendering for Overseas Work

In 1988 the world international market for building civil engineering and process engineering works totalled US $94,100 million (38). The six leading contracting nations: the USA, Japan, France, Italy, West Germany and the United Kingdom accounted for approximately 80% of this market. The United Kingdom held fifth place in the listing for overseas work. Whilst these figures did not isolate the total sum for civil engineering works alone, they indicate the significance of overseas works to the UK civil engineering industry.

Prior to the 1970's, UK contractors seeking work overseas had tended to work in the commonwealth and developing countries with which the UK had historical ties. During the 1970's, the rise in the price of oil, combined with the recession in the UK civil engineering industry, prompted civil engineering contractors to seek work in additional markets. Their attention was concentrated particularly in the Middle East and other oil producing countries, where oil reserves were being used to fund a rapid development of the countries' infrastructure. These countries have provided the UK civil engineering industry with substantial revenue from the construction of many large projects.

Many of the oil producing countries have now established their basic infrastructure thereby offering limited opportunities for further work. Other developing countries which previously provided work opportunities for UK contractors have subsequently encountered problems in the repayment of the large loans taken out to finance the works. The opportunities for further work in these countries are also limited and those projects that are programmed to be taking place are often subject to cancellation or delay. Now only some 15% of the UK contractors' overseas workload involves developing or oil producing countries. However, although the geographical location of overseas works may have changed, overseas markets remain highly important to the civil engineering contractor. The 1988 annual turnover for overseas works undertaken by the top 33 UK contracting companies totalled £1910 million (37). Western Europe was the most popular area for work followed by North America and the Far East. These areas were all listed above the Middle East in their importance to the contractors.

This chapter describes the main aspects of estimating and tendering for overseas works and provides examples of some of the main calculations undertaken. The format of the chapter follows that in Chapter 2 with the main differences between estimating and tendering for overseas works compared to those works in the contractor's home country highlighted.

THE ESTIMATING AND TENDERING PROCESS

An overview of the estimating and tendering process for overseas works is shown in Fig. 7.1. This shows the sequence of operations within the preparation of a tender for an overseas project and the main roles of the different functions within the contractor's organization. The comparison of this diagram with Fig. 2.1 shows that the processes involved with estimating and tendering for overseas works are basically the same as that for UK works. However, the type of project, the size of the project, the geographical location and the sources of relevant resource data all result in the need for additional parties to become involved with the preparation of the tender. As a result, the overall process becomes more detailed. This is reflected by an increase in the time allowed to the contractor for the submission of the tender.

The level of risk involved and the importance of the project to the company will determine the degree of involvement of senior management. The size of the project may require a team of estimators who will be managed by the Chief Estimator.

The construction of the works will be performed by the contractor under an agreed contract. In some countries local requirements will determine unique forms of contract applicable only to that country or particular project. Most usually however, the conditions of contract are based upon those of the Federation Internationale des Ingenieurs - Conseils (FIDIC). These conditions have been widely accepted by all the following construction industry organisations as a basis for construction (39):

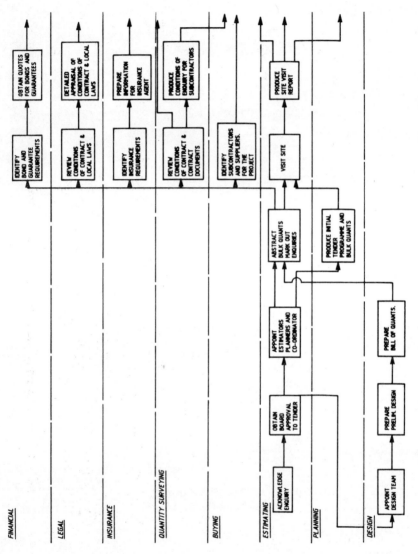

Fig. 7.1 The estimating and tendering process for overseas works

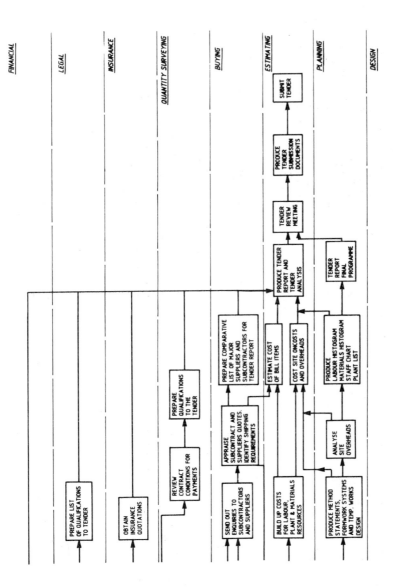

Fig. 7.1 The estimating and tendering process for overseas works (continued)

- Federation Internationale des Ingenieurs - Conseils;
- Federation Internationale Européene de la Construction;
- International Federation of Asian and Western Pacific Contractors Associations;
- La Federation Interamericana de la Industria de la Construccion; and
- The Associated General Contractors of America.

Under the FIDIC conditions the contractor is obliged to execute, complete and maintain the whole of the works. The conditions contain general clauses relating to both the definition and interpretation of the contract together with conditions relating to the particular project. In some instances the client will make amendments to the standard conditions with the deletion of standard clauses and the addition of extra clauses.

The contents of the standard FIDIC Conditions of Contract for Works of Civil Engineering Construction and their implications to the contractor are well documented in Jones (46), and other standard texts. It is èssential that the contractor fully reviews and understands the conditions applicable to the project under consideration as they will determine not only the contractor's contractual obligations but also the payment for work completed.

The personnel

The personnel involved in estimating and tendering for overseas work may be divided into the same three classes as before:
- the client's advisors;
- the contractor's personnel;   and
- external organisations.

The client's advisors
The client's advisors will be responsible for providing the contractor with the contract documents and dealing with specific queries. Whilst these tasks will normally be co-ordinated through one representative, the type and size of the project may well demand that direct access is provided by the client's advisors' staff to specialist consultants who have been involved with the production of the tender documents. The client's advisors will also arrange for the pre-bid briefing for the tendering companies and inform the contractor of any changes,

such as extensions of time, within the tendering conditions.

The construction contractor's personnel

All the construction contractor's personnel previously listed will play a role in the preparation of the tender:

- senior management;
- estimators;
- planners;
- buyers;
- plant managers;
- temporary works designers;
- in-house designers;   and
- site management staff.

In addition to the staff listed above, the preparation of a tender for an overseas project may require the involvement of the following staff:

- marketing staff;
- a tender co-ordinator;
- quantity surveyors;
- legal staff;
- the finance director;   and
- the personnel department.

The role of each of these additional staff is now examined.

Marketing staff

To secure contracts for overseas work contractors seldom rely solely on invitations to tender from foreign clients. The company will employ marketing staff to identify countries and geographical regions where future work is likely to be secured. In addition to the task of making direct contact with future clients, the marketing staff will assist estimating staff in the undertaking of territorial investigations. These investigations will accumulate information that is likely to be relevant if the company tenders for work in that location. This will include details on all aspects of commercial practice (such as labour laws and local taxation), local suppliers and contractors and competitors for work in the region. It is common for overseas clients to be aware of their project requirements but not to have reached the stage where funding is arranged and the design completed. In such cases the contractor may be willing to assist in the arrangement of finance and the design for the works. This work is often completed under a pre-tender agreement. Marketing staff play an active role in such agreements.

The tender co-ordinator

The function of the tender co-ordinator and the person who performs this task will be determined by the style of the contractor's organization and the staff currently available. The position within the tender team is often filled by a member of the construction staff awaiting assignment to a new project. It is the role of the tender co-ordinator to liaise between all the parties within the contractor's organization to ensure that information is collated and transferred to all those involved with the project. This is particularly important where there is a design element to the work because the impact of design decisions needs to be relayed to the estimating team.

It is the role of the tender co-ordinator to ensure that the contributions of all the parties involved are identified, prepared and included within the tender report. The tender co-ordinator will also keep in contact with the client's advisors to request any additional information required by the tender team. When this information is received from the client's advisors the tender co-ordinator will disperse the new information to the relevant personnel. Where the contractor considers that due to the non submission of specialist contractor quotations or other key information it may not be possible to complete the tender within the time allowed an extension of time may be requested through the Tender Co-ordinator.

Quantity surveyors

Whilst a contractor operating within his own country will normally be provided with a bill of quantities for the project, the documents provided by the engineer for an overseas project may not include a full measurement of the work. The contractor will then need to produce a bill of quantities type document to enable an accurate estimate to be produced. Where a bill of quantities has been supplied by the client's representative this may not be in the format normally recognised by the contractor and quantity surveyors will be needed to analyse and comment on the bill of quantities document. The contractor may suspect that there is an error in the quantities supplied by the client's advisors. In this situation the quantities will usually be checked.

Legal staff

The contractor's legal staff will normally be called upon to comment on the contract documents supplied by the engineer. Having been provided with the 'Conditions of Contract' they will be required to give early advice on the bonding requirements and any variations from Standard Conditions of Contract. During the tender period, the legal department will liaise with the quantity surveyors to ensure that enquiry letters to sub-contractors and suppliers (which will inevitably vary from standard conditions) are suitably worded. Before the final settlement meeting the legal staff will be required to review the offer letter and confirm any qualifications in the tender relating to legal aspects. If a Joint Venture is being proposed the legal staff will be responsible for the production of the Joint Venture Agreement.

The finance director

The estimator will, following the preliminary assessment of the estimate, provide the finance director with an assessment of the tender giving approximate values for the work divided into their constituents of local domestic and other relevant currencies. This ensures that the currency and exchange rate implications on the project may be investigated and the company's exposure to risk identified. The project bonding requirements will be given to the finance director so that these may be assessed and the appropriate bonding arrangements secured. The provision for local taxation is always an important factor in the finalization of the tender sum. This aspect will usually need investigation by the finance director who will advise on the correct allowance to be made when finalizing the tender.

The personnel department

The contractor's personnel department will provide the estimator with detailed information on the availability of staff and 'key personnel' required to man the project. This will involve the provision of staff conditions of employment needed by the estimator to calculate the cost of site supervision. Where new staff will need to be recruited, the availability, cost and lead time for recruitment will be given.

External organisations

In addition to materials suppliers, plant hire companies and sub-contractors, the production of a tender for an overseas project will require communication with:

- Local agents and representatives;
- Local construction companies;
- Local suppliers and sub-contractors;
- Shipping, packing and transportation companies;
- Banks and funding agencies;
- Joint venture companies;
- Principal sub-contractors;
- Risk managers and credit assessment companies;   and
- Local Embassy personnel.

During the tender preparation some or all of the above parties will need to be consulted. The parties that need to be contacted and the data gathered will depend upon the information already available to the estimating staff.

Local agents and representatives will be needed to advise on pricing factors including the availability of local labour, plant and materials.   They will also be able to give advice on shipping and distribution charges.

Local suppliers and sub-contractors will need to be contacted and their suitability for use on the project confirmed. The contractual basis for their involvement in the project will also need to be established.

Shipping, packing and transportation companies will need to be contacted to establish rates relating to the packing, handling and shipment of goods to the project site.

If the client has requested that the contractor assists in the raising of finance for the project, banks and funding agencies will need to be contacted and negotiations with respect to suitable finance packages commenced.

The contractor may decide that the project would be best suited by a joint venture arrangement with another offshore contractor or a local contractor. This will require extensive discussions at an early stage in the tender period to ensure that the parties concerned are able to establish the basis of a working relationship and have the expertise to meet the conditions of both the tender and the project.

Where the company has not previously undertaken work in the country, the territorial investigation made by the contractor is particularly important.

The contractor must be confident of the financial status of the client and other parties to the contract. If this information is not already available then credit assessment companies may be contacted for independent advice.

THE DECISION TO TENDER

The decision to tender for overseas work is clearly a major decision for the contractor. This decision is more difficult than that encountered when considering whether to tender for work within the contractor's own country. This is because frequently the work involved is for a client with whom the contractor has limited direct experience and the project may be located in a region where the contractor has little or no experience of previous work.

Against the potential profits must be weighed the cost of producing a tender, the likelihood of success and the risks involved with completing the project. In addition to the factors described in Chapter 2, which all affect this decision, the following main considerations are made by the contractor:

(i) What is the real likelihood of the project going ahead, and when?

(ii) Where is the project located?

(iii) Are we proficient in the construction of the work involved in the project?

(iv) Who is financing the project?

(v) How many competitors for the work are there?

(vi) Who are the competitors?

(vii) Do we have favourable experience of the client?

(viii) What is the duration of the project and does this justify the financial outlay in starting work in this new location?

(ix) Does the project offer adequate return for the risks involved?

The contractor must be reasonably confident that the project for which he has been invited to tender will actually go ahead when the tenders are received. Many large, prestigious projects exist in the minds of their promoters with very little probability of the project proceeding within a reasonable timescale. The contractor must establish whether finance has been obtained for the project, whether an external funding agency is involved and whether the project is within the country's structure plan and has received authority to proceed.

If the proposed project is located within a country where the contractor is currently employed on other projects, the contractor will have direct experience of working in that country and will probably have built up close links with local agents, materials suppliers, sub-contractors and other contacts in the locality.   This local knowledge should reduce the cost of tendering and the overall risk to the contractor of undertaking the work if the tender is successful.   Where the contractor has established a 'local office' in the country it may be necessary to tender for the project to maintain the workload in the region or to maintain the company's name on the tender list of the client.   Some UK contractors will only consider working in countries where English is the language of the contract.   Few contractors are willing to consider work on an ad-hoc basis in countries or regions where they have not previously decided to target their services.   If the contractor has had previous bad experience of working in a particular location or with particular clients then the decision may be made not to proceed with the tender.

Although the project may initially appear financially attractive most contractors act conservatively when faced with the opportunity to undertake work of which they have little previous knowledge and expertise.   Faced with such a project most contractors would either decline the offer to tender or look for a joint venture partner.

The contractor's major consideration at the decision to tender concerns the funding for the work to be undertaken and the contract conditions relating to payment for work completed.   The following questions are asked:
- Who will be financing the project?
- What will be the sources of funding?
- Are the contributors to the funding package committed?
- What are the consequences of the client reneguing on the payment of work completed? and
- What are the terms for payment in the conditions of contract?.

Before the contractor proceeds with the production of the tender and incurs the costs involved in the estimating and tendering process, he must be certain that, if successful, the project will both go ahead and result in regular payment to the company for the work completed.   If there is any doubt on this matter then the decision to proceed with the tender should not be made, or the contractor should seek some form of insurance against a possible loss in revenue after goods and services have been provided.

Insurance against defaults in payment or the imposition of unforeseen charges is not always available from private institutions.   Most countries provide some form of export credit guarantee through a national body.   In the UK insurance of export and overseas investment risks may be arranged through the Export Credits Guarantee Department (ECGD).   ECGD is a separate Government Department responsible to the Secretary of State for Trade and Industry and functions under the Export Guarantees and Overseas Investment Act 1978.   The Department provides support to exporters of goods and services to all types of industry, including the construction industry.   This support insures against the risk of not being paid via unconditional guarantees.   On this security banks are usually willing to provide finance to exporters.

The precise terms for the financial support vary according to the size and nature of the project and the country where the works are to be constructed.   Whilst the majority of overseas markets are covered occasionally the risks of non payment in a particular country are so high that cover is unavailable.   Similarly, some clients or projects may represent unacceptable risks.

This aspect of construction insurance is well described by Cox (44). The main risks to be covered are listed below:

(i)   The default of the government employer;

(ii)   Default or delay of the transfer of sterling payments;

(iii)   War between the employer's country and the UK;

(iv)   Revolution, war or large scale disturbances within the employer's country;

(v)   The imposition of import or export licensing or the cancellation of an existing licence for goods or materials;

(vi)   Additional handling, transport, or insurance charges due to the interruption or diversion of goods or materials being shipped from the UK;   and

(vii)   The employer's failure to pay the contractor sums awarded in arbitration proceedings under the contract.

Following the initial review of the contract documents the company must take a view as to the level of export credit guarantee required on the contract and seek appropriate assistance.   The premium is charged on the total of the estimated basic contract price and all additional sums within the project. There is usually a provision within the agreement for a proportional refund of a premium when the actual contract price falls short of the original estimate.

THE ESTIMATING PROCESS

(1)   Programming the estimate

The complexity of the estimating process when an overseas project is being considered, together with the number of parties involved make it imperative that a programme for the estimate is produced and then updated regularly throughout the preparation of the tender. This programme is used to identify clearly the roles of all the contractor's staff in the preparation of the estimate and the time scales allowed for the production of their appropriate contribution. It is particularly important that this programme shows the key dates of formal meetings within the tender period. It is essential that at the end of the tender period sufficient time is allowed for procuring the Bonds and Insurances because they are dependent on outside parties who will require an overall estimate of the project cost. The estimating team should keep closely to the estimate programme but is very dependent on the timely receipt of sub-contractor quotations. An example of an estimate programme is shown in Fig. 7.2.

(2)   The preliminary project study

As with any project, and particularly because of the risks involved with longer and more complex projects, it is important that the preliminary project study is undertaken with all relevant staff to establish the implications of the work involved. All the factors considered in Chapter 2, Section 2, must be considered. This will require expertise not only in the type of work involved in the project but in the staff who have knowledge of the geographical region and the client.

(3)   Materials and sub-contractor enquiries

(A)   Obtaining quotations for materials prices

The acquisition of materials prices for overseas projects is more time consuming, complex and costly than for similar projects in the contractor's

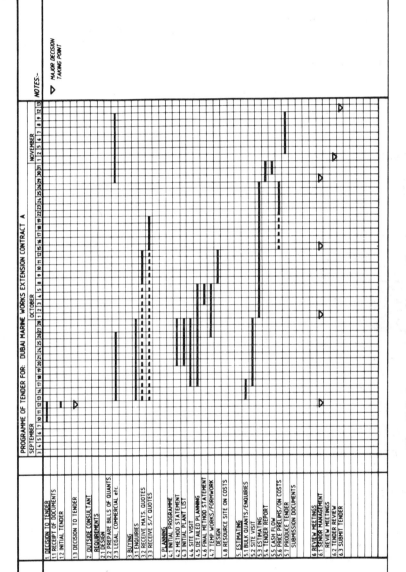

Fig. 7.2 An example of an estimate programme

home country.   Therefore adequate time within the tender period must be allowed for obtaining materials quotations.   Consideration must be made as to which materials may be obtained 'locally' (i.e. within the same country as the location of the project) and to which materials will have to be imported. In addition, resources such as water, not normally a problem in the contractor's own country, may present a particular difficulty in the country where the works are to be located.

Local materials

The contract documents may stipulate that certain materials for use in the project must be obtained locally.   Alternatively the contractor may choose to obtain certain materials locally in order to save on transportation costs.   The contractor's main concerns with respect to local materials relate to the quantity of material that will be available, whether the material will meet the specification required and whether the rate of delivery needed to meet the construction programme will be achieved.   A major part of the tender team's site visit will be checking local materials for use on the project and satisfying themselves that the demands of the project can be met.

Imported materials

Obtaining quotations for materials for overseas works is a specialist skill that requires a detailed knowledge of shipping procedures, transportation charges and shipping cost calculations.   This task is normally undertaken by buyers within the contractor's organisation specialising in overseas works.   The role of such a procurement department within the estimating process is to provide the estimator with prices for materials delivered to site.   This involves:

(i)   obtaining quotations for the materials items;

(ii)   providing freight, insurance, forwarding and clearance data for the materials and plant items to be used in the project;

(iii)   calculating freight and transport costs to site;

(iv)   investigating facilities for goods receipt and clearance, inland transport and all associated costs;   and

(v)   liaising with shipping agents, freight forwarders, airlines and all other parties regarding packing and shipping arrangements;

(vi)   checking that port handling equipment at the port of destination is adequate.

When considering suitable materials suppliers and sub-contractors for the project, reference must be made to the contract documents to check whether any restraints are laid down with the Instructions to Tenderers with respect to the 'eligible source countries' from which materials, equipment and services may be purchased. Depending on the location of the works and the source of funding, there may be constraints as to the source of goods and services that will be acceptable to the client.

Shipping cost calculations

For the materials that will need to be imported, the cost must be calculated of both purchasing the material and transporting it to the site. The material may be acquired anywhere in the world. The cost of the material on site must therefore include all freight charges, harbour dues, import duties, packing, insurance, etc. that will be necessary as the goods are transported from the country of origin to the site.

Any material to be shipped will have a weight and a volume. The shipping cost quoted by a freight agent will be calculated on the basis of the weight in 'freight tonnes'. This is equivalent to the actual weight of the material in tonnes or the volume of the material in cubic metres, whichever is the larger. The majority of goods shipped are in containers of standard dimensions. The procurement department will look at the option of using containers against alternative methods of shipping the goods and will make recommendations. The cost of shipping will be quoted in a specific currency, normally pounds sterling, US dollars or Deutschmarks at a specific exchange rate set at the time of shipment. Should currencies fluctuate significantly at the time of transport the shipper reserves the right to add a Currency Adjustment Factor (plus or minus), dependent on the strength or weakness of the quoted currency. Within the shipping contract there will also be provision for the shipper to charge additional monies because of increases in oil charges. This charge is commonly known as Bunker Surcharge. Allowance must be made in the estimate for additional costs due to Currency Adjustment Factors and Bunker Surcharge.

The price quoted by the supplier will usually be either the ex-works value of the goods or the charge of 'free-on-board' (FOB) at the port of departure. FOB represents the cost of the goods received over the ship's rail at a nominated departure point. Where the supplier has quoted an ex-works price, an allowance must be made for transport and insurance costs, together with any   other costs prior to loading on the ship.

To the freight cost must be added any forwarding agent's commission plus document charges. These charges will be dependent on the number of shipments and can vary considerably depending on the country to which the shipment is being made.   For example, for shipments to certain countries there is the requirement to obtain certificates of origin for the goods and legalisation costs to pay. Other countries require all shipping documentation to be translated into their language. This incurs translation charges.   All costs such as these are included within the documentation charge.

The cost of insuring the goods must also be included. When obtaining an insurance quotation an additional 10% is normally added to the carriage and freight cost of the goods to allow for inflation and any additional costs should the cargo need to be replaced. It is on this increased value therefore that the insurance quotation is based and the premium agreed. Once the material has arrived at the port of destination there will be additional costs for:

- import duties;
- local agent's commission;
- customs clearance;
- port handling;   and
- transport to site.

The estimate of all these charges will depend on the type of goods, the location and the local transportation problems.

Table 7.1 shows a typical cost calculation for the supply of polythene sheeting from the UK to a site in the Middle East. This calculation shows typical factors that need to be considered. As the cost build-up of diffeent materials varies, it is not a full and final analysis covering all materials.

Table 7.1   A typical shipping calculation for polythene sheeting to Dubai, Middle East

Supplier	=	ICI	
Shipping Specification	=	Weight	9900.00 kg
		Measure	16.00 m³
Basis of Costing	=	Measure	16.00
Shipping Currency	=	US Dollars	
Exchange Rate	=	1.67 US Dollars = £1	

Freight Rate	=	$134.85
Ex-works Value	=	Included
Charges up to FOB	=	Included
FOB Value	=	£10,071.00

Ocean Freight Charge	=	16.00 x $134.85	=	$2157.60
	=	$2157.60/1.67	=	£1291.98

(assuming 1.67 US dollars = 1.00 pound sterling)

Additional allowance for Currency Adjustment	=	Nil
Additional allowance for Adjustment in Oil Costs	=	Nil
Forwarding Agent's Commission	=	0.75% x £1291.98
	=	£9.69
Number of Shipments	=	1
Document Charges	=	1 x £190.00 = £190.00
Local Harbour Dues	=	Included within FOB cost

Gross Carriage and Freight Charges
= £1291.98 + Nil + Nil + £9.69 + £190.00
= £1491.67

Total Value of Goods including Carriage and Freight (C & F)
= £10,071.00 + £1491.67
= £11,562.67

Insurance Cost	=	0.50% x Total Value including Carriage and Freight plus 10% uplift
	=	0.50% x £11,562.67
	=	£57.81

Total Value including Carriage, Insurance and Freight, (CIF)
= £11,562.67 + £57.81
= £11,620.48

Import Duty	=	2.00% x Total Value including Carriage, Insurance and Freight
	=	2.00% x £11,620.48 = £232.41

Customs Clearance   =   £24.00 per consignment plus £2.67 per freight tonne

     =   £24.00 + (£2.67 x 16.00)

     =   £66.72

Port Handling   =   £3.28 per freight tonne

     =   £3.28 x 16.00

     =   £52.48

Transportation to Site   =   £6.89 per freight tonne

     =   £6.89 x 16.00

     =   £110.24

Local Agent's Commission   =   £2.50 per freight tonne

     =   £2.50 x 16.00

     =   £40.00

Total Local Charges   =   £232.41 + £66.72 + £52.48 + £110.24 + £40.00

     =   £501.85

Total Value of Goods including Carriage, Insurance and Freight delivered to site

     =   £11,620.48 + £501.85

     =   £12,122.33

The estimator will work closely with the buyer to ensure that the materials quotations are received at the earliest possible time for inclusion within the estimate. The buyer will usually seek quotations from different sources in different countries. These will then need to be compared and the appropriate supplier selected based on the final delivered to site cost. This is more than a simple cost comparison as it requires the difficult comparison of different performance specifications and the judgement of whether the materials proposed will meet the client's specification. Where the supplier is quoting for more than a single item then a detailed analysis must be prepared. The analysis is further complicated by the fact that the material may be purchased in one currency, transported by agents who demand payment in another currency and subject to local customs and transport charges in a third currency.

(B)   Obtaining sub-contractor quotations

To obtain sub-contractor quotations for overseas works requires attention to all the details as described in Section 3(B) of Chapter 2.   In addition, the contractor should ensure that the sub-contractor has taken into account all the factors related to the transportation of men and materials to site and that these are the same factors understood by the estimator to be the basis of the quotation.

The international nature of the work will mean that invariably sub-contractors from different countries are used in the project.   In addition to the usual information sought from sub-contractors detailed information on the sub-contractor's manpower resource levels will be required and adequate accommodation allowed for the sub-contractor's men.   Agreement must be reached as to who will supply this accommodation.

The importance to the project of the ability of the main contractor, specialist engineering contractor and sub-contractor to work effectively together is stressed by Cox (44), who states that this is normally the most important factor in the success of the overseas project.   Consequently the selection of sub-contractors should not be made solely on the rates submitted in their quotations but on their proven ability to work effectively with others on similar projects in similar geographical locations.

(4)   Project study, construction method and planning

As with construction in the contractor's own country, the planner and the estimator must work closely to gain a full appreciation of the work involved in the project.   This appreciation is gained by:
- a study of the contract drawings;
- an analysis of the bill of quantities and other contract documents;
- a site visit or country visit;   and
- the consideration of alternative methods of construction for the works.

The contract documents will be carefully studied. If the contractor considers that there are anomalies in the documentation or that information is missing, the client's advisers will be contacted for clarification.   Where necessary a bill of quantities document is prepared to produce a bill for estimating purposes.

For overseas work, a visit to the country where the works are located is required not only to gain as much information as possible about the project but also to update the territorial information already available to the estimating team.   The site visit should take place as soon as possible after the decision to tender and may be timed to coincide with the pre-bid briefing meeting with the client's advisors.   The geographical location of the site together with the time available within the tender preparation period will usually preclude more than one visit, so it is essential that all necessary information is obtained.   A schedule of site visit activities and check list of information should be prepared by the estimator in advance of the visit to ensure that all the required data are collected.

Pre-tender planning for overseas works should identify all the resources required to complete the construction.   All aspects of the project should be reviewed, and the resources required evaluated.   The planning team will:

- produce a full method statement and discuss the proposed methods with estimating and production staff;
- produce a programme for the works and discuss the programme with the relevant construction staff;
- produce a labour analysis and labour histograms;
- produce a plant list and issue the list to the plant department for them to obtain quotations, and to the procurement department for them to determine shipping costs;
- prepare a site layout showing offices, roads, plant installations, etc.;
- prepare management charts showing lines of responsibility and durations of staff employment;
- prepare a plant programme;   and
- prepare details of the contractor's general expenses and site on-costs.

The information collated may be incorporated into a 'Tender Planning File'.   This includes written elements and schedules to cover:

(1)   the site location

(2)   the site visit report

(3)   the bulk quantities

(4)   the method statement

(5)   the temporary works

(6)   the engineer's report

(7)   the formwork report

(8)   the scaffolding report

(9)   the management chart

(10)   the preliminaries build-up sheets

These sections will be cross-referenced as appropriate with drawings and graphical elements covering:

(1)   the site plan/location plan/photographs

(2)   selected architects' and engineers' drawings (e.g. floor plans, elevations and sections)

(3)   the site layout plan

(4)   the programme and labour histograms

(5)   the management chart

(5)   Calculating labour and plant costs

(A)   Labour costs

As with estimating for work within the contractor's own country all-in costs for labour should be calculated for each category of labour employed.   For overseas work, this requires consideration not only of the type of work to be undertaken by the labour but also the source of the labour.   Several possibilities exist:

-    local labour;

-    imported labour from the contractor's country of origin;   and

-    imported labour from one or more additional countries.

The contractor will normally identify the essential supervisory labour resources who will be employed to ensure that the project proceeds smoothly with the correct level of expertise.   Depending on the type of work to be undertaken, the remaining labour may be recruited via local agents or directly through agents from other countries.

The full cost of each of these labour categories must be reflected within the all-in labour rates calculated by the estimator.   Complicated calculations are necessary with different allowances made for different holiday conditions, air fares, medical, feeding, tools, clothing, non-working days and religious holidays.

The calculation of all-in labour rates is therefore even more important in overseas work and demands knowledge of all the different conditions applicable to different nationalities and religions.

(B) Plant costs

The calculation of plant costs for overseas works is extremely important in the production of the estimate. Unless suitable, reliable plant is known to be available for hire locally, the normal practice is to purchase new construction plant specifically for the project and export the plant items to the country concerned. The contractor's policy may be to allow for the full depreciation of the plant item over the project period and to treat any monies obtained from the sale of the plant item at the end of the work as an additional 'profit'. Alternatively, an estimate of the residual value of the plant item at the end of its period on site may be made and this allowance built into the calculation of the cost of the plant item per hour.

The fact that plant for the project is having to be purchased makes additional demands upon the estimator and the staff of the contractor's plant department. A detailed study must be made of:
-   the selection of correct plant item for the work;
-   the operator skills required;
-   the spares required to be held on site;
-   the tyre and fuel allowance for the machine;   and
-   the disposal of the machine.

The selection of the correct plant item for the work to be undertaken is clearly of great importance because of the capital cost involved. Whereas hired plant, if no longer required for the work, may be removed from site and the ongoing cost to the project ceased, an item of plant that has been purchased represents an ongoing financial commitment to the company. In overseas locations the plant item may not be easily sold if no longer required on the project. Therefore a detailed study of the plant requirements for the project should be made and these requirements carefully compared with the performance characteristics of the plant items. Wherever possible plant items suitable for different aspects of the construction should be selected.

Having identified the plant required, the estimator must be confident that suitable operators may be located and recruited for the project. The plant department will need to provide detailed information on the type and cost of spares which will need to be held on site for each plant item over the duration of the project. These items represent an additional cost which must be allowed for in the estimate. The high cost of tyres for wheeled plant requires study to be made by the estimator of the likely tyre requirements. Fuel requirements for the plant items will have to be estimated by consideration of the terrain and soil conditions of the site. If the estimator is making an allowance for the residual value of the plant item within his estimate then the likely market conditions and country of disposal for the plant item will have to be considered.

Every aspect of the acquisition, running and the maintenance of the plant item on the project (which may be a considerable distance from the normal plant distributor and maintenance depots) must be considered. The procurement department will need to establish the full cost of transporting the plant to site. Where the plant item will need to be assembled on site, an allowance both in time and cost for this work must be made. Where specialist items of plant (e.g. tunnel boring machines) are required, their fabrication and delivery must be carefully studied to ensure that their procurement complies with the requirements of the project programme.

The contract documents should be studied carefully to identify clauses relating to the importation and exportation of construction plant and equipment. The contractor may be required to pay customs duties relating to the import of materials and plant. These monies will be a combination of import duty, sales tax and surcharge fee which will vary with both the country concerned and the type of item. These duties are normally expressed as a percentage of the invoice value or as a certain rate against the materials quantity.

In some countries the amount of customs dues paid may be offset by a drawback scheme that incorporates a refund to the contractor depending upon the period that the temporary materials and plant have been employed within the country.

The preparation of a fully costed plant sheet

The planning team will normally prepare a schedule of plant for use on the project.   The estimator will then, in conjunction with the plant and procurement departments, build up a fully costed plant sheet to show the cost of the plant item to the project both in total cost and in unit cost per operational week or hour.   These data are then available for the estimator to use when pricing the bill of quantities for the work.   Where the plant costs are to be included in the preliminaries section of the bill this will be a direct transcription of costs.   Should it be desired to include the cost of the plant in the rates, the hourly rate would be input into the build-up, with a reconciliation being made to compare the monies built into the rates with the total plant cost.   An example of a plant sheet with costs included is shown in Fig. 7.3.

(6)   Estimating the direct costs

The estimator's task is to calculate the net cost to the contractor of executing the work defined in the contract documents.   This requires the calculation of all-in rates for labour and plant items together with obtaining prices for materials and sub-contracted work.   These are then combined with output rates or production times to determine direct cost rates for bill items or groups of bill items.   This process is shown diagrammatically in Fig. 2.2.

When the direct cost of construction work for overseas projects is estimated the same basic principles that have been described in Chapter 5 are applicable.   The calculations are however complicated by:

- the different currencies involved;
- the need to establish a tender exchange rate to be used in the calculations;
- the involvement of local and offshore monies;
- the chosen rates of productivity;   and
- the composition of construction gangs.

The Instructions to Tender will stipulate any conditions relating to the currencies for tender and payment for completed works.   For example, the following conditions relate to the tender for a typical civil engineering project in Pakistan (45), and indicate the typical constraints placed upon the estimator.

(1)   *The unit rates and prices shall be quoted by the Tenderer entirely in Pakistani Rupees.   A Tenderer expecting to incur expenditures in other*

PLANT ANALYSIS SHEET

Project:

Ref No.	Description	Units	Period on site in weeks	Hire rate per week	Total capital cost CIF	Total write off to contract	Total Spares	Tyres (Lump sum)	Period of use on site	Total fuel cost	Drivers total cost	Total cost to contract	Unit cost per week	Total residual value	% residual value
9011	Portable Compressor	4	76	N/A	28,800	23,040	3,456	2,676	70	6,400	Nil	35,572	117	5,760	20
9012	Lorry Mounted Concrete Pump	3	76	N/A	640,800	320,400	16,896	6,516	70	12,922	21,600	324,934	1922	320,400	50
9013	JCB 3X Hydraulic Excavator	2	76	N/A	57,600	38,592	6,912	7,523	70	15,601	18,980	87,608	576	19,008	33
9014	JCB 525 Fork Lift Truck	2	76	N/A	51,600	34,572	6,192	4,128	70	12,250	16,521	73,663	484	17,028	33
9015	3 Tonne Dumper	4	76	N/A	43,200	34,572	5,184	3,456	70	17,251	32,840	93,291	307	8,640	20
9016	150 mm Univac Pump	4	76	N/A	36,000	28,800	4,320	2,880	70	3,230	Nil	39,230	129	7,200	20
9017	25 Tonne Mobile Crane	2	76	N/A	307,200	153,600	16,864	3,720	70	16,153	21,640	211,977	526	153,600	50
	GRAND TOTAL	£ N/A	£ N/A	£ N/A	£ 1,165,200	£ 633,564	£ 59,824	£ 30,899	£ N/A	£ 83,807	£ 111,581	£ 866,275	£ N/A	£ 531,636	£ N/A

ALL SUM SHOWN IN POUNDS STERLING

Note: All figures shown above are for illustrative purposes only.

Fig. 7.3 An example of a plant analysis sheet

foreign currencies for inputs to the Works supplied from outside the Employer's country shall indicate the percentage of the Tender Price needed by him for the payment of such foreign currency requirements, either

(i)   *entirely in the currency of the Tenderer's home country, at the Tenderer's option; or*

(ii) *entirely in US dollars;*

always provided that a Tenderer expecting to incur expenditures in a currency or currencies other than stated in (i) or (ii) for a portion of the foreign currency requirements, and wishing to be paid accordingly, shall so indicate the percentage portion in his Tender. The amounts in various currencies, calculated on the basis of the percentages indicated in the Tender and by the use of the exchange rates indicated in the sub-clause (2) hereinafter, shall be used for the purposes of payment.

(2)   *The exchange rates to be used by the Tenderer for currency conversion shall be the selling rates for similar transactions published by the State Bank of Pakistan, prevailing on the date thirty (30) days prior to the latest date for the submission of Tenders. If exchange rates are not so published for certain currencies, the Tenderer shall state the exchange rates used and the source for the purpose of payments, the exchange rates used in the Tender preparation shall apply for the duration of the contract.*

It is therefore important that the contractor chooses a consistent currency for the calculation of the direct cost of the works, and also maintains a complete analysis of each different currency used throughout the calculations.

The normal practice for UK contractors is to undertake all the calculations in pounds sterling. The unit rates and prices within the tender submission are then converted to the required currency prior to the submission of the tender. These prices will be based upon a stated exchange rate.

It may be usual for the contractor when estimating for work in his own country to split the direct cost rate into the following cost categories:

-    labour;
-    plant;

- materials;
- sub-contractors;
- provisional sums;   and
- prime cost items.

For overseas works each of the above categories can be split into 'local' and 'offshore' rates. Additional categories for the rates may also be identified and decided upon as necessary. These important decisions and the exchange rates to be used in the calculations are normally decided during the Preliminary Project Study.

Particular attention is paid by the estimator to the identification of 'local' or on-shore monies (i.e. local currencies), and these are usually kept separate in the calculations. The implications of payments in local currency will be considered at the tender adjudication meeting, particularly if the local currency is a 'soft' currency, i.e. not exchangeable on the world financial market.

In order to produce the direct cost estimate the estimator must select appropriate output or production rates for the construction work. This will require consideration of the location of the project together with the gang composition and the productivity of the labour to be used. Within the same construction operation, the productivity rates assumed may be different, depending on whether the work is to be undertaken by ex-patriate, local or labour from other countries.

It is the calculation of the direct cost of the works that occupies the major part of the estimator's time within the tender period. In overseas works where invariably several estimators are involved in these calculations it is imperative that there is close liaison between the estimators to ensure that common and cost rates are applied throughout the tender.

(7)   On-costs

The on-costs for the project will be summarised and assessed under similar headings to those stated in Chapter 2, i.e.

(i) site management and supervision;

(ii) plant;

(iii) transport;

(iv) scaffolding

(v) miscellaneous labour;

(vi)   site office accommodation;

(vii) temporary works and services;

(viii) general items;

(ix)   commissioning and handover;   and

(x)   sundry requirements.

The checklists provided in Tables 5.50 to 5.55 for UK works may be used as an indication of the details to be included.

In addition to the on-costs normally included when preparing a tender for a project in the contractor's own country, consideration must be given to:

- additional staff costs;

- accommodation for staff and labour;   and

- bonds, guarantees and insurances.

Additional staff costs

The estimator must identify which supervisory and clerical staff will be from the contractor's own personnel and will be employed on the project with expatriate status. Some supervisory and clerical staff will need to be recruited locally or from another country. This will be dependent on the skills available and the local labour laws. Table 7.2 lists some of the considerations within the employment package of the UK supervisory staff for an overseas project.

Accommodation for staff and labour

Depending on the size and location of the project the contractor may need to provide accommodation for site management, supervisory staff and the labour required for the project. This will normally take the form of housing for expatriate staff and a labour camp for other workers. The facilities provided will depend very much on the location and the country of origin of the work force.

Bonds and guarantees

The contractor will have additional on-costs for those bonds, guarantees and insurances not normally encountered when tendering for works within the contractor's own country. Under English law, a 'bond' is a guarantee in which one party binds himself (or his executors or assignees) to pay a stipulated sum to another party (or their executors or assignees). There are two types of such bond or guarantee: unconditional and conditional. An unconditional bond is

Table 7.2   Employment considerations for
UK supervisory staff overseas

Leave allowance in days   :   Married staff/Single staff

UK leave air fares   :   Married staff/Single staff

Local leave air fares   :   Married staff/Single staff

Schooling allowance in UK   (£/annum)

Schooling allowance (local)   (£/annum)

Terminal bonus   (as a percentage of basic salary)

Baggage allowances (in kg)   Single

                                 Married

                                 Wife

                                 Children

Pension allowance on UK equivalent salary   (£ sum)

Staff replacement

Estimate of staff requiring replacement during the duration of the project

                                 (percentage)

Salary conditions

UK equivalent salaries   (£/annum)

Percentage paid on shore in local currency (percentage)

---

one which may be called for at any point in time whenever the beneficiary decides. A conditional bond is dependent upon certain conditions or events taking place.

In the procurement of overseas contracts, bonds, guarantees and insurances represent a considerable cost to the contractor. The cost of these items is based upon the final estimated project cost. The main types of bonding needed to be obtained by the contractor are as follows:

(a)   Tender bond;

(b)   Performance bond;

(c)   Advance payment bond; and

(d)   Retention bond.

These different types of bonds are described below.

(a)   Tender bond

With few exceptions, the tender must be accompanied by some form of security as a guarantee that the contractor, if declared the successful tenderer, will enter into a contract with the employer for the amount of the bid sum and will provide the necessary contract bonds. Tender bonds (also known as bid bonds or proposal bonds) are widely used for the purpose of bid security, although some clients may require that each contractor submits a certified cheque or other form of negotiable security. When the contractor becomes the successful tenderer the bond is retained until such time as the contract is signed and satisfactory contract bonds are provided. If the contractor is unsuccessful in his tender his tender bond will normally be returned when the bid is formally rejected.

The value of the tender bond will normally be between 5 and 10% of the total tender sum. The bond will normally be denominated in the currency of the tender or the currency of the country where the project will be located. The tender bond will be 'called' if the tenderer withdraws his tender or, in the case of a successful tenderer, fails to sign the contract and furnish a performance bond in accordance with the conditions of contract. Although tender bonds only represent a small percentage of the project cost, the cost to the company of the provision of such bonds across many concurrent tenders soon represents a significant overhead cost to the company.

(b)   Performance bond

Under the terms of the FIDIC Conditions of Contract (39), the contractor must obtain, when required by the employer, a bond or guarantee of an insurance company, bank or other approved sureties for the performance of the contract. The terms of the bond or guarantee must be approved by the employer. The insurance company, bank or other party providing the surety must be approved by the employer and the party severally bound with the contractor to the employer for the due performance of the contract. The costs in securing the bond must be met by the contractor. Failure of the contractor to meet this fundamental contract requirement justifies the employer in terminating the contract.

If awarded the contract, the contractor must normally, within thirty days of the receipt of notification of the Intent of Award from the client, furnish a performance bond. The face value of the bond is derived from a percentage of the total contract price. The bond guarantees the performance, completion and maintenance of the works. In the USA the performance bond is known as a surety bond and is designed to ensure that the client does not suffer if the contractor fails to meet his obligations. A surety bond may be up to 100% of the value of the contract. The cost of the bond must be borne by the contractor and as such represents a significant on-cost to the project. The premium rate for the bond will typically vary with the type of project and the project duration. A sliding scale normally applies with a rate quoted across different ranges of contract value. Premiums for performance bonds are normally paid annually in advance and are reduced as the contractor completes the work.

(c)   Advance payment bonds

Where the client is providing an advance payment to the contractor to cover the mobilization for the contract or site preparation this payment will normally be secured by an advanced payment bond. This bond protects the client from non completion of the preparatory work. The amount of the bond may reduce as the monies are spent and as the mobilisation on site occurs. The amounts of the bonds vary with the level of the advanced payment and the period over which the advance payment is repaid.

(d)   Retention bond

Retention bonds may be used in place of the normal retention sums. The amount of the bond, as with a normal retention sum, is normally between 5 and 10% of the total value of the work undertaken by the contractor. The client may agree to the release of retention monies to the contractor subject to the provision of a retention bond. If the contractor is able to negotiate this arrangement, the release of these monies will reduce the capital that the contractor has 'locked-up' within the project.

In order to secure bonds from an appropriate financial institution the estimator must provide sufficient information relating to the project and precise details of the bonding or type of guarantee required. Where the contracting company is part of a larger group of companies the bond may be guaranteed by the parent company who will be capable of obtaining appropriate cover at preferential rates.

Insurance

Under the FIDIC Conditions of Contract (39) the contractor is required to obtain insurance cover for:

- care of the works   (Clause 20);
- the insurance of the works   (Clause 21);
- third party liability   (Clause 22);
- third party insurance   (Clause 23);
- liability for accident or injury to workmen   (Clause 24);
- interference with traffic or adjoining premises   (Clause 29).

It is essential that the contractor obtains an all risks policy which, with certain exceptions, will provide the necessary insurance cover to meet the works to be erected and all materials, plant and other items which belong to or are the responsibility of the contractor during the project period.

Insurance cover for plant and materials required for use on the project and needing to be transported to site will normally be obtained under a marine or overland insurance policy. To obtain such insurance it is important that the estimator is prepared to supply some or all of the following information as soon as it becomes available.

(i) A brief description of the works.

(ii) Details of any joint venture agreements relating to the project and a clear statement of the company's involvement and the controlling party for insurance arrangements.

(iii) The location of the project.

(iv) Details of the conditions of contract and any amendments from standard conditions.

(v) An estimate of the contract value divided into direct works, subcontracts, preliminaries, etc.

(vi) The contract period and maintenance period.

(vii) An estimate of the total value of construction plant and equipment together with the maximum value at any one time and the period of risk.

(viii) Details of any temporary works included in the project.

(ix) Details of the expatriate accommodation and all temporary site accommodation and offices.

(x) Site plans and general arrangement drawings for the project.

(xi) Information relating to climatic conditions, tidal conditions (where applicable) and difficult ground conditions (e.g. high water table, solid rock, loose sand, etc.).

(xii) Details of excavations deeper than 5 m.

(xiii) Details of any superstructures with foundations/basements/ podiums/towers/dams/reservoirs/seaworks/tunnels/pipe-lines/ bridges.

(xiv) An estimate of the number, type and value of all motor vehicles to be used on site.

The insurance of the works and other major items within the project will normally be arranged within the contractor's own country. The insurance of other items not included within the FIDIC conditions of contract, for example site transport and staff motor vehicles, will normally be arranged locally in the country where the project is located.

The collation and management of on-costs

How the estimator collates and recovers on-costs within the tender price will depend upon the tender documents and the preferred method of apportioning these costs. For overseas projects on-costs and other general overheads form a high proportion of the total project value. The calculation of on-costs and their inclusion within the tender submission requires careful consideration. As with the inclusion of on-costs within a tender for work in the contractor's own country, it is important that the contractor fully recovers these on-costs. Their recovery should not be at risk if the client decides to change the quantities of work within the project.

THE TENDERING PROCESS

The tendering process for overseas works can be divided into four stages:

(1) The assessment of the estimate and the evaluation of adjustments;

(2) The assessment of general overheads;

(3) The assessment of risk and profit;   and

(4) Tender submission.

These stages are the same as described in Chapter 2 when considering work in the contractor's own country and involve the same calculations as described in Chapter 6.

The assessment of the estimate and the evaluation of adjustments

Because of the complexity of overseas projects and the need to make bonding arrangements, some contractors have two tender settlement meetings. The first, chaired by the estimating director and attended by certain members of the tendering team and construction staff, is primarily to review and agree the method to be employed in construction together with the indirect resources. This enables the estimator to finalise the direct cost estimate for the works and obtain bonding requirements.

The second settlement meeting is chaired by the managing director and is attended by the estimating director plus all members of the tender preparation team. The contribution of other specialist professionals such as quantity surveyors, legal and financial staff would normally be by written presentation. At this meeting key aspects of the tender will be reviewed. These may include reference to:

- the fundamental level of pricing;
- labour and material resource availability;
- the outputs used in calculations;
- indirect costs;
- design fees;
- insurances and bonds;
- overheads and profit;
- risks;
- provision for management fees and taxation;
- cash flow and the need for funding;
- the capital expenditure and residual value of plant;   and
- the tender letter and any qualifications.

The assessment of general overheads

The project under consideration must make a contribution to the general overheads of the company. This contribution will be dependent upon the company's head office costs and anticipated turnover. Most projects have project co-ordination and support costs. For certain projects, particularly those involving a design element, it may be necessary to extend the head office facilities or staffing levels to meet the demands of the project. In such cases an additional allowance towards head office costs must be included within the tender.

The assessment of risk and profit

The additions for profit and risk will be based upon the same factors as described in Chapter 6. The risk element for overseas works is much greater and therefore requires more consideration than the risks for a similar project in the contractor's own country. It has been suggested that the allowance for risk made by a contractor working overseas is often the largest single item of cost in his tender (43).

Risks may be divided into:
- contractual risks;
- client risks;
- construction risks; and
- economic risks.

Within these four categories the following main issues will need to be reviewed:

- climatic factors, (e.g. hurricane, earthquake, etc.) unique to the location of the project;
- the availability and quality of the labour;
- the contractual conditions (particularly payment terms);
- the role of joint venture partners and major sub-contractors;
- materials availability and suppliers; and
- possible currency fluctuations.

Where the project is to be constructed under a fixed price contract, careful consideration must be given to the impact of inflation and changes within the principal quantities of work.

Risk is assessed by experience and the latest information available on all the other parties involved in the project. In some situations the views of outside risk assessors may be sought to provide expert opinions on the type of project, the client and the location of the project.

Tender submission

Where the client has provided a bill of quantities document to the contractor the item rates and totals will be entered into the bill for submission. The information submitted by the contractor to the client for overseas works tends to be more comprehensive than for work in the contractor's own country. The tender submission will depend upon the requirements stated by the client's advisors within the tender documents. The submission may include all or some of the following documents:

- details of the joint venture structure;
- a labour schedule;
- a plant schedule;
- a priced bill of quantities;
- a schedule of stage payments;
- the form of tender;
- a tender programme;
- a method statement;
- an organisation chart;
- the required bonds;
- the tender letter;
- details of any finance offer;   and
- alternative designs.

Any qualifications to the tender will be included within the tender letter. Some client's advisors will require that all the documents are marked with the seal of the contractor's company and dated. The conditions of tendering may demand that the tender documents supplied to the contractor are returned with the tender submission.

Having submitted the tender, the contractor awaits feedback from the client's advisors as to the acceptability of his bid. When tendering for overseas works the response from the client is seldom immediately affirmative, and the contractor may become involved in detailed negotiations surrounding the tender submission before the tender is finally accepted by the client.

The cost of estimating and tendering for civil engineering works overseas is often considerable. Within some countries there are grants available from government sources to help support the cost to the contractor. Even with such financial support the decision to obtain overseas works and establish a reputation for undertaking construction on an international scale represents a major decision for any civil engineering contractor. Whilst the return may be high, the risks are also high as are the costs of estimating and tendering, often with little initial success. Nevertheless, overseas works represent an attractive market in which some companies continue to reap significant financial benefits.

8   Estimating and Bills of Quantities

INTRODUCTION

The procedure whereby clients or promoters of construction, both building and civil engineering, employ or engage professional representatives to design the works and then, after receiving and comparing tenders, select a contractor to construct the works is common practice. Although in the fullness of time it may change, this practice seems set to continue and dominate for a very long time. Examining ways of coping with the consequences is therefore still a worthwhile occupation.

It is through the contract documents - drawings, specification and bill of quantities - that the professional representatives of the promoter of engineering projects describe the works to be constructed. The measurement contract based on a bill of quantities is the most common type of contract in use in the UK. The bill of quantities is the contract document that has had the most influence on the practices and procedures of contractors' estimators. The main reason for this is that the bill of quantities establishes the data required by the promoter in the tender submission. To provide the data required by the tender documents is a major requirement of contractors' estimating and tendering procedures. While the principal task of the estimator is to produce an estimate of cost, to his company, of constructing the works, the way in which the estimator does this is clearly affected by the need to satisfy the requirements of the tender documents, particularly the bill of quantities.

Changes in the method of measurement or the format of the bill of quantities have consequential effects on the working procedures of estimators. The replacement of the ICE Standard Method of Measurement by CESMM in 1976 and the second edition CESMM2 in 1985 and the replacement of the

Standard Method of Measurement of Building Works, SMM6, by SMM7 in 1988 are the more recent examples. These changes introduced a more structured and codified presentation of measurement information.

Changes in the work practices of estimators do not cause the method of measurement to alter, except by influencing the development of the next major revision, and contractors' estimators seeking to evolve their working procedures are largely constrained to do so within the current framework of contract documents and procedures. Such is the situation, current at the time of writing. Many contractors have adopted computer based estimating methods and many are either in the process of adopting these new methods or are considering their adoption. The reason for contractors undertaking these changes is typically, *inter alia*, to make better use of the limited pool of experienced estimators, to use 'in-house' company data more effectively and to provide a more structured framework for the direction and supervision of the estimating personnel. These reasons are not influenced at all by the contract documents or the format of the bill of quantities but the computer based methods have to cope with the existing contract documents.

CURRENT METHODS OF MEASUREMENT

The main methods of measurement used in civil engineering works are:

(1) The Institution of Civil Engineers:
    'Civil Engineering Standard Method of Measurement', second edition, 1985 (CESMM2) (5).
(2) The Department of Transport 'Method of Measurement for Highway Works' 1987 (6).

In building works, the current method of measurement is the 'Standard Method of Measurement for Building Works', seventh edition, 1988, SMM7 (17), although, at the time of writing, use is still made of SMM6 (16).

Both CESMM2 and SMM7 provide a working structured classification with notation which could be used by estimators for storing estimating data as well as for referencing bill items. There are no current known proposals for replacing the DOT's 'Method of Measurement for Highway Works' by one incorporating a structure classification with notation.

ESTIMATING

Estimators use a variety of techniques when preparing an estimate, the main ones being operational estimating, unit rate estimating, spot rate estimating and sub-contractor quotations as described in earlier chapters. Computer aided estimating methods have sought to provide all the techniques that the estimator uses manually and to support the estimating methods with 'filed' or 'stored' data wherever possible and helpful to do so.

Operational estimating is favoured by the civil engineering estimator because it is more appropriate for plant intensive work where allowances for 'down' time, 'travel' time and 'idle' time can be more readily included and because it links well with the planning based approach to estimating used by civil engineering contractors, the elapsed times or durations being derived from the plans or programmes. The use of filed data to support the estimator is largely limited to single contracts, with the plant and labour 'configuration' for an operation (such as excavation) being uniquely derived on each occasion. The civil engineering estimator also uses unit rate estimating widely for work not estimated by the operational method, the main reason being its convenience for producing rates for the items in bills of quantities. The use of unit rate estimating is also prevalent in building work, the reasons being the greater proportion of items not involving heavy plant and again the convenience of the method for producing rates to be set against bill items.

With unit rate estimating the use of filed data can be extensive and the facility to store a build-up or even a range of build-ups for an item of work can readily be developed. The difficulty lies in retaining a convenient link between the bill of quantities and the stored data. This difficulty is one of establishing a link between a bill item description and the stored build-ups. This difficulty is seen with both civil engineering and builders' estimating.

COMPUTER AIDED ESTIMATING AND METHODS OF MEASUREMENT

Many contractors are now introducing computer aided estimating systems. (An example of such systems is described in Chapter 9.) The majority of these systems allow estimators to store data that may be used to price bill items.

The task of linking bill item descriptions to filed estimator's data is, *prima facie*, easiest with methods of measurement such as CESMM2 or

SMM7 because of their structured classification and notation.   Taking CESMM2, for example, and extracting a bill item from a real bill with the description

>Formwork type C. fair finish
>
>Plane vertical to plinths to pumps and blowers
>
>width not exceeding 0.1m

is under the CESMM2 given the item reference or the code G241.   This bill item can thus be uniquely identified by the reference:

>Bill Number/Section Number/Page Number/Code, viz.
>
>1 / 1 / 1 / G241

Against the code G241 the estimator's unit rate build-up can be stored. The link between the bill description and filed data is thus established through the code.   In any of a variety of available estimating systems such data are stored in a library file known as the 'performance data file', 'the make-up file', 'the build-up file' or whatever name is allocated the file of unit rate build-ups. These build-ups contain the labour, plant and materials resources together with their output or usage rates.   The costs of these resources can be taken from created resource cost files which can be managed separately.   Thus in the build-up file the CESMM2 code, in this case G241, is used to identify the data required by the estimator.   The fact that G241 is also the item reference is helpful in that the estimator in identifying this reference does not need to translate the bill description into a 'code'.   This has already been done in creating the bill description by using the CESMM2 code.

There are difficulties which arise, one being that the estimator requires several different build-ups to be available for each basic bill item description. For example, the labour and plant used to excavate small areas of topsoil would be quite different to the labour and plant used to excavate large areas of topsoil.   Thus the configuration of labour and plant varies with quantity, whereas the basic bill item description remains the same.   Another example would be trench excavation where the output of the excavation team would vary with ground conditions whereas the underlying bill item description is substantially the same.   To meet these circumstances there needs to be in the estimator's reference an additional 'character' which allows for 'choice of method'.   In some systems the first decimal place is used for this purpose, thus:

>G241.0 to G241.9

gives up to ten options on the choice of estimator's build-up for the same basic bill item description. This creates a difference between the bill item reference and the estimator's reference since the decimal point is used in the item reference for varying the description. The use by the bill author of the decimal place is not usually sufficiently important to warrant a different estimator's reference. This difference between the item reference and the estimator's reference, creating possibly two references per item, is in fact essential to accommodate a second difficulty which is that of the bill item reference being miscoded by the bill author. Thus the creation of the facility whereby a different code refers to the item and the estimator's library of build-ups overcomes the two problems of offering the estimator choice of method and accommodating mis-codings.

Fig. 8.1 shows a page extracted from a bill of quantities prepared using the CESMM2. The estimator's references have been added. This shows that in most cases the item reference and the estimator's library reference are the same, the exceptions being a few cases where the decimal place has been used by the estimator to exercise a choice on the method of construction as represented by the build-up stored on file.

Figs 8.2 and 8.3 show the build-ups stored against the estimator's references used for item F411.1 and E311.0.

In building work similar opportunities to use the item reference as a link to the estimator's data are available under the classification and notation systems of SMM7. For example the following bill item 'Formwork to sides of upstands, plain vertical height 250-500 mm' is under SMM7 given the reference code E20:4.1.3.0. This bill item can be uniquely identified by the reference

bill number/section number/page number/code viz.

1 / 1 / 1 /E20:4.1.3.0

Against the code E20:4.1.3.0 the estimator's unit rate build-up can be stored. The two problems are (a) having a variety of build-ups against a single code and (b) bill authors miscoding item references still remain and thus encourage separate coding systems.

Thus for items that estimators wish to cost using computers and data held on computer files, the approach is either:

(i)   Mark the bill with the library reference codes as shown in Fig. 8.1 and have the data preparation staff enter this code. Having this

	HEADWALLS, CULVERTS AND SEWERS TO EXISTING BYE-PASS BRIDGE					
	In-situ Concrete					
	Provision of concrete; prescribed mix for ordinary concete taken from Table 50 of C.P. 110: Part 1					
F213	Grade 10; 20 mm aggregate	)	m³	8	F213	.2
F214	Grade 10; 40 mm aggregate	)	m³	52	F214	.2
F233	Grade 20; 20 mm aggregate	)	m³	1	F243	.2
F243	Grade 25; 20 mm aggregate	)	m³	41	F253	.2
	Placing of Concrete					
	Mass					
F611.1	Blinding 75 mm thick	)	m³	1	F411	.9
F611.2	Blinding 75 mm thick; to slope	)	m³	7	F411	.9
F624	Bases, footings and ground slabs over 300 mm thick	))	m³	52	F423	.8
	Reinforced					
F722	Bases, footings and ground slabs 150 – 300 mm thick	))	m³	31	F522	.1
F742	Walls 150 – 300 mm thick	)	m³	7	F542	.1
F773	Casting to metal sections (pile up) 0.1 – 0.25 m² cross sectional area	))	m³	3	F573	.4
	Concrete Ancillaries					
	Formwork; rough finish					
G143	Plane vertical width 0.2 – 0.4 m	)	m²	15	G143	.2
G144	Plane vertical width 0.4 – 1.22 m	)	m²	23	G144	.1
G145	Plane vertical width over 1.22 m	)	m²	21	G145	.1
	Carried to Collection					

Fig. 8.1   Page from a CESMM2 bill with estimator's library reference marked

Work Group F411.1

PLACING OF MASS CONC. BLINDING N.E.150 mm (CRANE AND SKIP)

Units of Measurement m³

Code	Description	Cost/Hour	Usage	Cost/m³
L2	LABOURER (CONC)	£5.26	1.500	£7.89

Code	Description	Cost/Hour	Output	Cost/m³
P5	MOBILE CRANE	£27.00	9.000	£3.00
P99	SKIP	£2.50	9.000	£0.28

Total labour cost/m³	£ 7.89
Total material cost/m³	£ 3.28
Total net cost/m³	£11.17

Fig. 8.2   An estimator's build-up stored in the library against code F411.1

Work Group E311.0

EXCAVATE TOPSOIL N.E.0.25 m

Units of Measurement m³

Code	Description	Cost/Hour	Usage	Cost/m³
L101	LABOUR	£5.10	0.125	£0.64

Code	Description	Cost/Hour	Output	Cost/m³
P12	CAT 951 EXCAVATOR	£27.00	12.000	£2.25
P205	16 TONNE TIPPER	£25.00	4.000	£6.25

Total labour cost/m³	£ 0.64
Total material cost/m³	£ 8.50
Total net cost/m³	£ 9.14

Fig. 8.3   An estimator's build-up stored in the library against code E311.0

code associated with the bill item there is a link between the bill item and the estimator's stored build-up.   An automatic link through to the cost of the resources an 'estimate' for this is now in place in the computer system.   The estimator is left to check or inspect the implied estimate.

or

(ii)   Enter the library reference code at a later stage when the estimator himself is operating the computer system.

In either case the result is that the build-up stored on file becomes attached to the bill item.   The estimator is able to edit this build-up to render it unique for a particular contract against code E311.0.

In CESMM2 and SMM7 bills the link between bill items and the estimator's unit rate build-up held in his or her library file can be achieved through the bill item reference code derived from the method of measurement. This has drawbacks which have been described and which lead to coding systems which are independent of the reference code.   Neither the DOT 'Method of Measurement for Highway Works' nor the SMM6, which is still in use, presents such opportunities and structured classifications with notations suitable for use by estimators have to be established, frequently by individual companies.

CLASSIFICATION AND NOTATION

Theories of classification have mainly been developed for use in library systems.   The user's contact with a classification system is through the notation.

The purpose of establishing notations for bill items is to store and retrieve estimators' data.   Sher (56) established criteria which a notation system for estimators' data would need to satisfy.   These criteria were:

(a)   Unique identification
Each item in the bill needs to be identified by a unique and unambiguous reference.

(b)   Length of notation
The length of notation should be as short as possible.   There is a conflict between length of notation which, when short, it is argued reduces errors in

entering notation and reduces time in entering notation but increases time taken searching unless stratified. On balance the shorter notation has more benefits.

(c)   Recognisability of notation

The user of the classification should easily be able to translate an item in a bill of quantities into its notation and vice versa. With a well designed notation it should be possible more or less to guarantee to get the notation right first time so that one can check it at a glance. This aspect is a function of the presentation of the classification. Stewart (87) identified several factors which contribute to good design when investigating the manner in which information is displayed on visual display units. It is argued that these are: logical sequencing, spaciousness, relevance, consistency, grouping and simplicity.

(i)   Logical sequencing: The sequence in which the notation is presented should be logical in terms of the user's task.

(ii)   Spaciousness: Spacing and blanks in the presentation of the notation are important, both to emphasise and maintain the logical sequencing or structure and also to aid the identification and recognition of items of information. Clutter greatly increases search time and increases the likelihood of missing or overlooking items, misreading items, and other errors.

(iii)   Relevance: There is a natural desire to ensure that the user has available all information of real and potential relevance. In many cases, information of only potential relevance should be excluded. This aspect is dealt with in more detail later.

(iv)   Consistency: The value of consistency is that an unfamiliar section can be more readily and accurately interpreted if it conforms to existing practices in its use of language and structure.

(v)   Grouping: Where there are relationships between items it can improve the presentation if relevant items are grouped together.

(vi)   Simplicity: All the above factors should be taken into account, but the overriding consideration should be to present the appropriate quantity and level of information in the notation in the simplest way.

(d)   Level of detail

It is not envisaged that a notation for estimators' data would cater for the multitude of different items that may occur in a bill of quantities. The sheer volume of items would make this impracticable. What is required is that any notation should cover the items of work normally undertaken by the organisation in question. In other words a contractor specialising in factories will develop a data library relating to that building type, whereas another contractor with a more varied workload will require a more general range of items. The degree to which a notation should be refined to cater for more unusual items depends upon the organisation in question.

(e)   Description of materials

The inclusion of the numerous different materials which may be used in construction in any classification and notation would seem to be impracticable and impossible and would require an inordinate amount of computer storage. The system should distinguish between materials that make differing demands upon other resources. For example, the cost of pouring concrete mix A or concrete mix B will not vary greatly in any given location. However, any difference in reinforcement content may have significant consequences for the cost of the placed concrete. The materials should be able to be allocated on an ad hoc basis into broad categories incorporated in the classification.

(f)   Accommodation of dimensions

In order to store performance data relevant to various items, the classification must allow for the definition of dimensions as and when required.

(g)   Expandability

The requirement of expandability is referred to by all authorities in the subject of classification and notation so that existing established notations do not need to be changed.

(h)   Option for choice of method

When estimating an item, the estimator's choice of resources depends upon many factors. These include the quantity of work involved, restrictions (if any) on working conditions and the height of the work above ground level. If the estimator is to use an estimating system he has to be able to store and retrieve performance data peculiar to differing conditions. To enable him to do this, the classification should in some way represent the required method of operation.

Notation for use with DOT method of measurement and SMM6

Using the criteria described above as guidelines and also the DOT library of standard item descriptions (7) as a skeleton a notation was developed for the DOT 'Method of Measurement for Highway Works'. Similarly a notation was developed for the SMM6. This work was undertaken by the Department of Civil Engineering at Loughborough University of Technology in order to support the development of computer aided estimating within a number of construction companies. The procedure for development was:

(i)   To create a notation for each section of the Method of Measurement;

(ii)   To test this notation by coding bill items from existing bills;

(iii)   To issue the notation section by section to all company estimators involved;

(iv)   To have the estimator's code bill items from existing bills;

(v)   To receive the estimator's comments;

(vi)   To revise the notation;   and

(vii)   To repeat the steps (ii) and (vi) until a consensus emerged.

Fig. 8.4 gives a sample page from the notation developed for use with the DOT 'Method of Measurement for Highway Works'.

Fig. 8.5 gives a sample page from the notation developed for use with the SMM6.

Fig. 8.6 gives a sample item build-up stored against a reference from the DOT Method of Measurement notation.

Fig. 8.7 gives a sample build-up using the SMM6 notation.

H. Sub-base and Pavement				
1. Sub-base	A. Granular Type 1	1. In carriageway	A. 60 mm thick	
2. Roadbase	B. Granular Type 2	2. In emergency crossg	B. 70 mm thick	
3. Lower roadbase	C. Cement bound mat. Cat.1	3. In layby and bus bay	C. 80 mm thick	
4. Upper roadbase	D. Cement bound mat. Cat.2	4.	D. 90 mm thick	
5. Base course	E. Cement bound mat. Cat.3	5.	E. 100 mm thick	
6. Wearing course	F. Cement bound mat. Cat.4	6.	F. 110 mm thick	
7. Pavement	G. Wet lean concrete 1		G. 120 mm thick	
	H. Wet lean concrete 2		H. 130 mm thick	
	I. Wet lean concrete 3		I. 140 mm thick	
	J. Wet lean concrete 4		J. 150 mm thick	
	K. Wet mix macadam		K. 160 mm thick	
	L. Dense tarmacadam		L. 170 mm thick	
	M. Dense bitumen macadam		M. 180 mm thick	
	N. Rolled asphalt		N. 190 mm thick	
	O. Bitumen macadam		O. 200 mm thick	
	P. Tarmacadam		P. 210 mm thick	
	Q. Dense tar surfacing			
	R. Cold asphalt			
	S. Open textured bit. mac.			
	T. Open textured tarmacadam			
	U. Jointed RFCD concrete			
	V. Unreinforced concrete			
	W. Continuously RFCD conc.			
	X. 10 mm AGG			
	Y. 14 mm AGG			
	Z. 20 mm AGG			
	AA. 25 mm AGG			
	BB. 40 mm AGG			

SECTION 7 PAVEMENTS

Fig. 8.4 Sample page from the notation developed for use with the DOT 'Method of Measurement for Highways Works'

G. REINFORCEMENT	BARS	(TONNE)	STRAIGHT AND BENT BARS
	1. MILD STEEL DIAMETER 6 mm	A. IN FOUNDATIONS	A. HORIZONAL 12.00 TO 15.00 m
	2. DITTO 8 mm	B. IN GROUND SLABS	B. DITTO 15.00 TO 18.00 m
	3. DITTO 10 mm	C. IN SUSPENDED SLABS	C.
	4. DITTO 12 mm	D. IN WALLS	D.
	5. DITTO 16 mm	E. IN CASING TO STEEL COLUMNS	E. VERTICAL 5.00 TO 8.00 m
	6. DITTO 20 mm	F. IN CASING TO STEEL BEAMS	F. DITTO 8.00 TO 11.00 m
	7. DITTO 25 mm	G. IN CASING TO STEEL COLUMNS AND BEAMS	G.
	8. DITTO 32 mm	H. IN STEPS	H.
	9. DITTO 40 mm	I. IN STAIRCASES AND STRINGS	CURVED BARS
	10.	J. IN STAIRCASES AND STRINGS AND ASSOCIATED LANDINGS	I. HORIZONTAL 12.00 TO 15.00 m
	11.	K. IN STEPS, STAIRCASES, STRINGS AND ASSOCIATED LANDINGS	J.
	12.	L. IN TOPS OF DORMERS	K. VERTICAL 5.00 TO 8.00 m
	13. HIGH YIELD DIAMETER 6 mm	M. IN TOPS AND CHEEKS OF DORMERS	L.
	14. DITTO 8 mm	N. IN MACHINE AND SUNDRY BASES	M. LINKS, STIRRUPS, BINDERS AND SPECIAL SPACERS
	15. DITTO 10 mm	O. IN ISOLATED COLUMNS	
	16. DITTO 12 mm	P. IN ISOLATED BEAMS AND LINTELS	
	17. DITTO 16 mm	Q. IN ISOLATED COLUMNS, BEAMS AND LINTELS	
	18. DITTO 20 mm		
	19. DITTO 25 mm		
	20. DITTO 32 mm		
	21. DITTO 40 mm		
	22.		
	23.		
	24.		

Fig. 8.5 Sample page from the notation developed for use with SMM6

Work Group 03A9G.0

FABRIC REINFORCEMENT BS REF B785

Units of Measurement m²

Code	Description	Cost/Hour	Usage		Cost/m²
L31	STEELFIXER	£6.30	.300		£1.89

Code	Description	Cost/Hour	Usage	Wastage	Cost/m²
M55	MESH B.S.REF B785	£3.6/m²	1.000	20.0%	£4.32

Total labour cost/m²	£1.89
Total material cost/m²	£4.32
Total net cost/m²	£6.21

Fig. 8.6 Sample item build-up stored against a reference from the DOT Method of Measurement notation

Work Group HA100H.0

FMWK TO BATRNG. END OF WALL-275 mm

Units of Measurement m

Code	Description	Cost/Hour	Usage	Uses		Cost/m
LG38	CARP GANG(FWK)(6+1) (M+F+S)	£18.99	1.500	1.0		£28.48

Code	Description	Cost/Hour	Usage	Uses	Wastage	Cost/m
XG1	SFTWD/UNLD/NAILS	£253.40/m³	0.012	1.0	6.0%	£3.20
XG2	18 MM PLY/UNLD/NAILS	£6.59/m²	0.275	1.0	5.0%	£1.90

Total labour cost/m	£28.48
Total material cost/m	£ 5.12
Total net cost/m	£33.60

Fig. 8.7 Sample item build-up stored against a reference from the SMM6 notation

Using the notations developed

The notation is used in the same way as the CESMM2 codes described earlier, the main difference being that the item references are not the same or similar to the estimator's library references.   Fig. 8.8 shows a page from a bill prepared under the DOT Method of Measurement with the notation or codes superimposed.   These can be input as with the CESMM2 as follows:

(1)   The estimator (or his assistant) has the option of writing notations against each bill item or not.   If notations are provided at this stage, the data preparation staff will key in the notation together with the bill reference. This establishes a link between the bill item and the estimator's library of data and the bill item is linked to the data stored and referenced by that notation.

(2)   If the estimator does not code the bill as above, the option exists at a later stage for the estimator to provide the notation when considering the individual bill items.

By either action (1) or (2) above the estimator can link the bill item to filed build-ups.

A point worth recording is that if the estimator operated through the data preparation staff as in (1), then there are two possible sources of error:

(a) The estimator provides the wrong notation;   and

(b) The data preparation staff key in the notation incorrectly.

The safeguard against both of these is that the response displayed on the VDU is headed with words describing the item, and these words are held on file.   This allows the estimator to determine whether this was the build-up that he intended to use.   The quality of the system in use is frequently determined by the quality of the safeguards incorporated.

Creating an estimator's data library

Having established the notation which will be used to identify the estimator's data and link them with work items or bill item descriptions the estimating department has to provide the data that will accompany each notation.   The notations used, depending on the company, can be the CESMM2 codes, SMM7 codes, those for the DOT Method of Measurement or SMM6, or one specially created by the company itself.   Having the notations or codes available the estimators must decide:

(a) The trades or classes (e.g. excavation, concrete work, etc.) for which a data library will be created;

(b) Within the trade or classes which specific work items will be included in the library;   and

(c) Which work items will have multiple build-ups representing different construction methods.

This will produce for the estimators a target list of work items, extracted from the list of codes or notation.   For example, using the CESMM2 codes, the estimators may decide to include Class F: IN SITU CONCRETE, in the data library.   They may also decide to include all the provision of 'normal' concrete in the data library which would be

'F111.0 to F388.0'

Furthermore the estimators may decide to include three methods on provision, one providing small to medium quantities of concrete using a mixer, one providing large quantities using a large batching plant, and one using a build-up which simply uses ready mixed concrete.   Thus the estimators have created a catalogue of work items for which they are prepared to create a data library. One example is:

F232 Provision of concrete, designed mix, grade C20, cement to BS12, 14 mm aggregate.

The estimator is going to create three build-ups for these items:

F232.1 using a small concrete mixer;

F232.2 using a batching plant;   and

F232.3 using ready mixed concrete.

This exercise would be repeated throughout each of the major trades or classes.

The task is now to write down the build-up for each of these work items.   Fig. 8.9 shows a standard proforma for this task.   Fig. 8.10 shows this complete for the build-up that is displayed in Fig. 8.2.   At the time these proformas are completed a second table of resources needs to be created.   This is a list of individual resources or gangs or groups of resources that are contained in the build-ups being recorded on the standard proformas.   For example, labour may be given the codes L1 to L2.   Thus the estimators must decide how many categories of labour they will use.   There are frequently only three:

Number	Item Description	Unit	Quantity	Rate	Amount £	p
60.	600mm x 900mm x 50mm thick precast concrete paving flags in paved area on Granular Type 1 sub-base 100mm thick to surface sloping at more than 10° to the horizontal.	sq.m	253.00		—	
	SECTION No.14 – FORMWORK FOR STRUCTURES					
61.	Formwork more than 300mm wide horizontal or at any inclination up to and including 5° to the horizontal – Class F3.	sq.m	8.80		N1Z3A2.0	
62.	Formwork more than 300mm wide at any inclination more than 5° up to and including 85° to the horizontal – Class F1.	sq.m	8.20		N1Z1A1.0	
63.	Formwork more than 300mm wide at any inclination more than 85° up to and including 90° to the horizontal – Class F1.	sq.m	1169.83		N1Z1A1.0	
64.	Formwork more than 300mm wide at any inclination more than 85° up to and including 90° to the horizontal – Class F3.	sq.m	31.06		N1Z3A1.0	
65.	Formwork more than 300mm wide at any inclination more than 85° up to and including 90° to the horizontal – Class F4.	sq.m	17.96		N1Z4A1.0	
66.	Formwork 300mm wide or less at any inclination – Class F1.	sq.m	13.92		N1Z1B.0	
67.	Formwork 300mm wide or less at any inclination – Class F3.	sq.m	5.68		N1Z3B.0	
	SECTION No.15 – STEEL REINFORCEMENT FOR STRUCTURES					
68.	High yield steel bar reinforcement nominal size 16mm and under of 12m length or less.	tonne	32.924		O1B1.0	
69.	High yield steel bar reinforcement nominal size 20mm and over of 12m length or less.	tonne	24.851		O2B1.0	
	SECTION No.16 – CONCRETE FOR STRUCTURES					
70.	Insitu concrete Class E in blinding 75mm or less in thickness.	cu.m	16.39		P2A.0	
70.	Insitu concrete Class 30/20.	cu.m	679.00		P1L.0	
				Page total		

Fig. 8.8   Sample page from a DOT bill with reference notation added

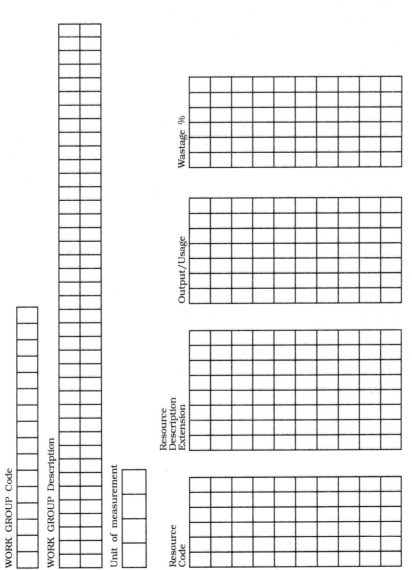

Fig. 8.9 Standard proforma for assembling build-ups

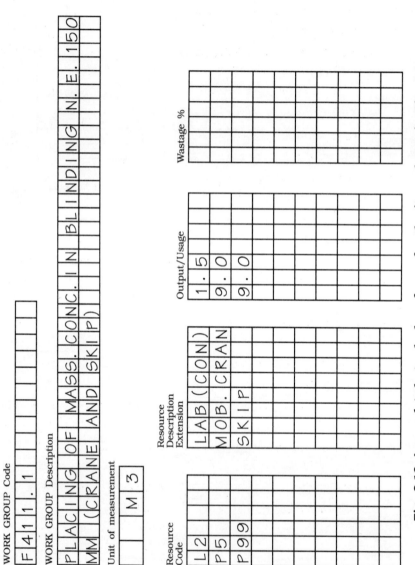

Fig. 8.10 A completed standard proforma for the item shown in Fig. 8.2

L1   LABOURER

L2   CRAFTSMAN

L3   PLANT OPERATOR

However, it could be more and carpenters and fixers etc. could be identified separately with their own codes, viz.:

L1   LABOURER

L2   CARPENTER'S LABOURER

L3   CONCRETE LABOURER

L4   STEEL FIXER'S LABOURER

L5   CARPENTER

L6   STEEL FIXER

L7   PIPE LAYERS,   etc.

This allows the creation of gangs, viz.:

LG1   CARPENTER GANG

     L5 - CARPENTER x 3

     L2 - CARPENTER'S LABOURER x 1.

Similarly materials may be given codes M1 to M2 and materials groups the codes MG1 to MG2, plant may be P1 to P2 and PG1 to PG2. Mixed groups involving labour, plant and materials may be given the codes XG1 to XG2. If in creating a build-up the estimator includes, say,

L2   LABOURER

P1   22 RB TRACKED CRANE

M99   READY MIXED CONCRETE

then he must ensure that these resources are recorded in the list of resources.

The following shows the sequence of events for creating a data library.

ACTIVITY	RESPONSIBILITY
1. Decide which classes of work are to be included.	Chief Estimator
2. Decide which work items within the classes to include.	Chief Estimator
3. Decide which work items are to have multiple build-ups.	Chief Estimator

4.	Write out proformas creating build-ups.	Estimators
5.	Create list of resources from build-ups.	Estimators
6.	Enter resources and build-ups into the system.	Data preparation staff
7.	Check entered data.	Estimators

9   Computer Aided Estimating

Throughout the last decade there has been a great increase in the use of computers and associated information technology within civil engineering contractors' organisations. The introduction of the micro-computer brought computing to all departments within contractors' organisations. Previously, the use of computers had been limited to the large and medium sized companies who, via mainframe and mini computers, concentrated on the processing of payroll and accounting information. The cheap, personal processing power of the micro-computer has provided the platform for the development of applications software packages to meet the needs of engineers and management throughout contractors' organisations. Software suppliers, recognising the size of the civil engineering and building market, responded accordingly and the result has been the development of software for all types of construction management functions.

In 1987 a survey entitled 'Building on IT' was conducted by the Construction Industry Computing Association (CICA), in conjunction with KPMG Peat Marwick McLintock (71). This survey contacted over 800 organisations representing the largest consultants and contractors in the building and civil engineering industry. From the respondents, detailed information was collected on their current and anticipated use of information technology. In 1989 the survey was repeated and the results published in the report 'Building on IT - for the 90's' (72). Fig. 9.1 shows a summary of the data collected on the use of computers by contractors, indicating the respondents' figures for the first year of mainframe, mini and micro-computer use. The figure shows clearly the impact of computing into contractors' organisations during the 1980s.

Table 9.1 shows the survey's results for the use of computer software

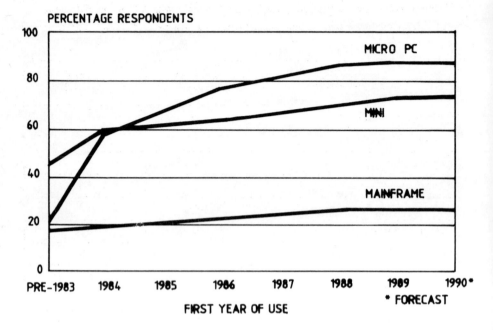

PERCENTAGE RESPONDENTS

Fig. 9.1   Contractors' use of mainframe, mini and micro-computers (first year of use) (from CICA/KPMG Peat Marwick McLintock Survey, 'Building on IT - for the 90's')

by civil engineering contractors.   As expected, accounting was the top application together with financial modelling using spreadsheets. The results of the survey indicated that contractors are cautious when considering the implementation of systems to assist with technical applications and there has been a guarded approach to such applications as estimating, valuations and plant management (72). Computer aided estimating was used by some fifty percent of civil engineering contractors.

This level was confirmed in a survey undertaken by the Construction Management Group at Loughborough University of Technology (73).   This survey of the members of the Federation of Civil Engineering Contractors (FCEC) was conducted to establish the current use of computer aided estimating systems and their impact on the tasks of the civil engineering estimator.

Table 9.2 shows the use of computer aided estimating subdivided into companies with small, medium and large annual turnover.

Table 9.1 Civil engineering contractors' use of computer software (this table has been prepared using data from CICA/KPMG Peat Marwick McLintock Survey, 'Building on IT - for the 90's')

APPLICATION AREA	Current Use %	Under Consideration %
Word Processing	78	85
Desk Top Publishing	21	35
Spreadsheets	93	93
Databases	42	56
Management Systems	55	64
Project Management	50	50
Contract Costing	70	70
Order Processing	35	55
Estimating	57	77
Valuations	42	57
Plant Management	36	50
Computer Aided Draughting	21	35

Table 9.2 Use of computer aided estimating by small, medium and large companies

Do you use a computer aided estimating system?

Response	Number of responses by annual volume			Total
	Small	Medium	Large	
Yes	3 (16%)	23 (62%)	7 (81%)	43 (56%)
No	16 (84%)	14 (38%)	4 (19%)	34 (44%)
No response	0 (0%)	0 (0%)	0 (0%)	0 (0%)
Total	19	37	21	77

The growth in computer usage for estimating has been matched with the number of commercially available computer aided estimating packages. In 1979 there were some 12 packages available on the UK market (58), by 1985 the number had risen to 80 (70).   Currently the market has stabilised below this figure.   Some products are no longer available and other products have been acquired by larger software suppliers and combined with their existing range of products.   This specialist area of construction computing now has established market leaders.

Whilst many companies have undertaken the implementation of computer aided estimating packages, it would be incorrect to infer that all those companies who have attempted to implement computer aided estimating systems have done so successfully.   Computer aided estimating remains one of the most difficult application areas with respect to the selection, testing and implementation of software systems (74).   Some companies have attempted to introduce computer aided estimating systems only to find the software unsatisfactory for their particular organisation. Other companies who elected to develop their own systems have discovered the trials and pitfalls of software development.   Nevertheless, there are many contractors who have successfully introduced computer aided estimating systems within their organisations. This has been achieved by careful software selection and planned progressive implementation.   These systems are now fully operational and producing both foreseen and unanticipated benefits to the companies concerned.

ESTIMATOR'S REQUIREMENTS IN COMPUTER AIDED ESTIMATING

Estimator's requirements in a computer aided estimating system relate primarily to the features available within the system.

As part of a research programme in the use of computers in estimating at Loughborough University of Technology the following facilities were identified as being fundamental to a computer aided estimating system.   A computer aided estimating system must:

- calculate bill item prices from input data by a number of different methods;
- apply calculated item rates against all relevant bill items;
- provide an extension and summation of bill item prices to produce direct cost totals;

- provide a variety of reports and bill listings for the estimator and other company personnel within the estimating and tendering process;
- store data on different resources and their requirements for different construction methods;
- store lists of all-in rates and materials and sub-contractor prices for the contract under consideration;
- store the full build-up of each bill item within the contract with the facility to retrieve, check and rework the item if required;
- assist the estimator in his communication with other parties both inside and outside the contractor's organization;
- enhance estimators' skills and extend their knowledge of the construction processes;   and
- limit potential errors within the estimating process.

Software developers and suppliers have in the main recognised these requirements and developed systems to meet the majority if not all of the facilities listed above.   The logical and easily definable nature of these requirements means that, from a computer system's viewpoint, the requirements may be easily met. It is the user interface, i.e. how the estimator uses the system, and the speed of operation that present the main barriers to computer aided estimating.   The system may contain all the functions required but when called upon in an everyday environment it may be found to be unworkable. For example, the entry of bill item data prior to the build-up of an item cost may be too time consuming to justify the adoption of the system.   Compared with manual methods, the editing of data may be too cumbersome for estimators to adopt.   Where the method of system operation is satisfactory the speed of the system may simply be too slow to meet the relatively short time period available to produce the tender.   Consequently, systems that initially appear attractive may be found to be unacceptable in practice.

DIFFERENT TYPES OF COMPUTER AIDED ESTIMATING SYSTEMS

The term 'computer aided estimating' covers a range of different types of systems developed to meet different needs in different types of companies. For the purpose of this text, the following main types of computer aided estimating system will be considered:

- systems based on spreadsheets;

- bespoke systems; and

- estimating packages.

These types of computer aided estimating systems represent the main use of computers for estimating purposes. Other systems, such as those based upon planning packages, are not usually suited to pricing bills of quantities documents. Consequently their use is limited to those situations where the estimator is not required to submit a priced bill of quantities as part of the tender.

The use of spreadsheets for computer aided estimating

The growth in the availability and use of software products such as spreadsheets and databases has produced a range of computer aided estimating systems with facilities to support the estimator in his work. Such facilities have usually been produced by estimators to meet their own individual requirements.

Electronic spreadsheets are one of the most ingenious software developments and the first electronic spreadsheet, launched in 1979, revolutionised the micro-computer industry. It was probably the major single influence on the growth of micro-computer usage by commercial organisations.

A spreadsheet is a two dimensional grid made up of horizontal rows and vertical columns with each intersection of a row and a column forming a cell. This cell may store data and is identified by its unique address designated by the row and column reference. The data held in the cell may be a number, a label (descriptive text) or a formula (an instruction for performing calculations with other cells). Spreadsheet commands may be used to operate on the data stored in the cells as well as to change the appearance of the spreadsheets. These commands are operated through a simple user interface making it easy for relatively inexperienced computer users to produce simple systems to perform routine calculations. The main attractions of the spreadsheet are the ease with which calculations may be edited and repeated and the spreadsheet extended to accommodate additional data.

The early spreadsheets provided only simple mathematical functions within the command structure. The development of spreadsheets has produced a current range of products that have become complex software programming tools. Amongst the features provided are:

- the ability to calculate in 'three dimensions' across a range of spreadsheets;
- the ability to link to other software packages and products; and
- the ability to perform almost any mathematical operation such as frequency distribution, multiplication of matrices, linear regression, and the computation of the present value of a series of future cash flows.

These features, together with the use of 'macros' (a set of instructions made up of a sequence of key strokes and commands) enable the user's work session to be automated, removing the necessity for the repetitive keyboard work. Spreadsheets are consequently suitable for assisting estimators in their calculations.

The early use of spreadsheets for estimating purposes was usually limited to the production of calculation sheets for determining back-up calculations such as the 'all-in' rate for labour, the total cost of preliminary items or the cash flow for the project. The use of spreadsheets has now been extended into the area of organising and maintaining data on resource costs and outputs.

Although some spreadsheet programs are available commercially and publications providing the format for estimating related calculations may be purchased, there do not appear to be any commercially available estimating systems based on spreadsheets. However, the popularity of spreadsheets amongst computer users will ensure the continued development of the products and it is anticipated that their use by estimators will continue to increase.

Bespoke estimating systems

A bespoke system is one that is developed specifically for the users within one particular organisation. The development may be undertaken totally 'in-house', i.e. within the contractor's organisation, or the software may be produced in conjunction with a specialised software supplier. The need for such a development is normally the result of the company being unable to locate a suitable application package that will meet both their estimating and computing requirements.

The development of a bespoke system has a number of main advantages:

(i)   because the system is produced specifically for the estimators within the company, the system includes all the functionality required for their work;

(ii)   the estimators have an opportunity to participate in the development of 'their' system (this makes the acceptance of the system at the implementation stage less of a problem); and

(iii) the company is normally able to ensure the quality of software that is developed and that comprehensive testing is undertaken prior to the introduction of the system.

The main disadvantages are as follows:

(i)   the cost of the system production and maintenance is considerably higher than a packaged solution;

(ii)   there is normally a very long lead time between the start and completion of the system; and

(iii) the production of the system is only possible with the full commitment of management and users at every level within the organisation.

As with any bespoke software development, the development of an acceptable computer aided estimating system is dependent on the production of a comprehensive Functional Specification Document that identifies in exact detail the requirements of the users. This requires that the users have an understanding of what the computer system is capable of producing for them. It is essential that, before a bespoke system is produced, the users are educated to the extent that they are capable of recognising the full potential of the system and the requirements of the system designers and developers. Conversely the production of a satisfactory system is also dependent on the knowledge of the system's designers in the relevant application area. Only with this knowledge can the design team fully interrogate the users to identify their true requirements and design fully acceptable systems.

The time and cost involved in the production of a bespoke system mean that they are usually only produced by the large construction companies. Their development requires strong project management and the total commitment of all those concerned with the project. Too often unsatisfactory systems are produced after considerable time, money and energy has been expended. However, the bespoke solution remains the only sure way of producing a system that exactly meets the needs of users. Projects such as the development of BBEST, an estimating system produced under a Teaching Company Scheme by Balfour Beatty Construction Ltd in conjunction with the Department of Civil Engineering at Loughborough University of Technology, shows that with adequate resources and proper management a totally acceptable and subsequently working system can be produced. Research on the development of other systems has shown that the cost and lead times of bespoke systems can be reduced through the application of systems analysis methodologies and the utilization of fourth generation languages (67). As with the BBEST system, extensive use of prototyping should be made to ensure that users clearly understand the implications of the design proposals. There is no substitute for 'What you see is what you get'.

Estimating packages

There are currently numerous commercially available estimating packages. These provide a range of features and facilities and may be 'purchased' at a range of different prices. The 'Construction Industry Software Selector', published annually by the Construction Industry Computer Association, provides a full listing of the packages available to the industry (75).

Whilst different packages offer different features the majority provide the same basic facilities;

- A data library for storing data or resources and construction activities;
- A range of methods for pricing the items within a bill of quantities;
- The ability to enter project specific data;
- The facility to amend the data on a project at any time and produce a revised estimate for the project;
- A range of methods for adding additional monies to the direct cost estimate to produce a tender sum; and
- Comprehensive reporting facilities to enable the estimator to produce summary and detailed information from the estimate.

The main advantage of estimating packages is their cost compared to bespoke systems development. The purchaser may, by careful selection procedures, be reasonably confident that on the acquisition of the system a comprehensive complete estimating system will be available for immediate use.

The disadvantages of application packages are as follows:

(i)   It is unlikely that the system available will fully meet the purchaser's requirements. A decision has to be made as to how critical the unavailability of specific features is to the potential users.

(ii)   The system installed by the supplier cannot be amended by the user.

(iii)   Suppliers are normally reluctant to customise their software for particular users. The user is dependent upon the next 'upgrade' of the system to include the additional functions required.

(iv)   The contractor is totally dependent on the supplier to maintain the software and ensure that the system is functioning satisfactorily.

(v)   If the supplier ceases to trade or the business is acquired by another party who is reluctant to meet the supplier's commitments the users may find that their investment is unprotected.

Notwithstanding these disadvantages the purchase of an application package remains the best method of acquiring a comprehensive proven estimating system at a reasonable cost. As with the acquisition of any product it is sensible to select a package from an established supplier who has a proven track record of producing satisfactory software. The best guide to the standard of the product and the service supplied is the supplier's existing customers, who should be contacted before a system is purchased.   If no such customers exist, a thorough testing or trial of the system is essential.

HARDWARE DEVELOPMENTS FOR USE WITH COMPUTER AIDED ESTIMATING

System developers are continuing to develop better products to meet the needs of practising estimators. This includes the development and customisation of additional hardware for use with computer aided estimating systems.   The development of computers and associated information technology continues to provide better data processing, handling and storage facilities.   A comprehensive computer aided estimating system may, in addition to the computer system on which it is based, include one or all of the following additional hardware elements:

- an optical character reader;
- a mouse; and
- a digitizer.

Optical character reader

An optical character reader (OCR) is the name given to a hardware device that provides the facility to scan a page of text and identify the characters present. The characters may then be processed via the computer system to enable the text to be edited or stored directly within the system. The use of OCR's with estimating systems is described by Morris (68) and (69).

The majority of OCR's use either hardware or software methods of character recognition. The hardware technique is termed 'matrix matching' and involves dividing the page into a grid pattern and then 'reading' by a series of photo-electric cells. The grid is compared with a stored set of printer fonts to identify the most likely character. The software method of a character recognition involves 'feature analysis' and examines the shape of the character that has been read in terms of various characteristics, e.g. vertical lines, horizontal lines, loops, diagonals, etc. These characteristics are then compared with known characters to identify the most likely character and font.

The level of character recognition achieved when using an OCR is dependent on the type of OCR and the accuracy of the system. After a page of text has been read by an OCR it is possible to correct errors with the use of editing facilities. However, if too many errors exist, the time taken to make the corrections outweighs the advantages of using the OCR to input data. Given the wide variety of type faces used in the production of bills of quantities and the quality of some bills forwarded to the estimator, the difficulty in using OCR's for computer aided estimating is clearly apparent.

Significant developments are however being made which are improving both the speed and accuracy of character recognition. As the technology improves, their use by estimators for computer aided estimating will increase because of the need to enter the data from the bill of quantities document as fast as possible.

Mouse

A mouse is an input device consisting of a mobile hand held unit connected by a cable to the computer system. The unit is capable of transversing a flat smooth desk top and, by either electromechnical or optical means, transmitting information to a computer. Provided the software has the correct 'program' to recognise the information, the signals may be transmitted into cursor movements. As the mouse is moved across the desk top the cursor on the screen moves accordingly. The mouse unit will be fitted with one of several different button type arrangements. These buttons are used to select menu options within the application program. This provides the user with an alternative interface to the standard keyboard. By adapting their systems to operate with this form of input device, software developers are providing additional flexibility in data input and system usage. This method of input is preferred by many novice computer users who find the use of a keyboard difficult.

Digitizer system

A digitizer system consists of an electronic or magnetic tablet which is used with a transmitting pen connected by a cable to the computer. The digitizer tablet may be supplied in a number of different sizes up to a size large enough to accommodate an A0 drawing. When the scale of the tablet's dimensions have been set by the user, the system may be used to input data direct from the drawing by touching the appropriate points on the drawing with the tip of the pen. The digitizer is an input device particularly suited to measuring dimensions directly from construction drawings and is invaluable for use with computer aided estimating systems. If the estimator is supplied with a digitizer, bill measurement system and a computer aided estimating system, dimensions of the work items within the project may be measured directly from the drawings, the totals calculated and then passed directly to the estimating system, so reducing the amount of data entered via the keyboard to a minimum.

Computer aided estimating systems which incorporate these hardware features are now commercially available. One which makes use of all three is the 'Conquest' system. The main features and facilities within the Conquest system are now described.

The Conquest computer aided estimating system

The Conquest computer aided estimating system has been developed by Digital Building Systems Ltd. Digital Building Systems have been at the forefront of computer aided estimating for the last decade and have systems operational in some 600 construction companies. Conquest shows the combination of the latest technology with extensive experience of how estimators work. A fast efficient user interface has been developed which makes computer aided estimating both effective and efficient.

Conquest overcomes the main barriers to computer aided estimating which to date have made estimators reluctant to take up computer based methods. The system is flexible, fast and easy to use with the facilities to input data from a number of different methods. Conquest reflects the way in which estimators price bill items manually and represents the 'state of the art' of commercially available computer aided estimating packages. All the principal facilities required within a computer aided estimating system are available:

- data libraries;
- flexibility in pricing bill items;
- easy amendment of project data;
- comprehensive reporting; and
- full mark-up facilities.

These are accessed via an efficient, fast, user-friendly interface that removes the need for complex coding of project and library data. Conquest represents a new generation of computer aided estimating systems that should enable more estimators to adopt computer based estimating methods. An outline of the system is described here to show how, over the last decade, systems developers have understood estimators' requirements and worked to meet their needs.

The Conquest system user interface

The Conquest system incorporates the latest software techniques to provide the estimator with a number of different ways to select the data required and the operation to be performed on these data. The system allows the estimator to input data via a keyboard, digitizer or an optical character reader. Once data have been entered, they may be selected by:

- displayed lists of menu commands;
- a menu of commands displayed horizontally at the bottom of the screen (a ring menu);
- menu options direct from the digitizer board; or
- highlighting key fields on the screen and indicating that this is the field on which the estimator wishes the system to function.

This selection may be made via the keyboard or with a mouse attached by cable to the computer.

The system makes full use of the keyboard facilities with the use of cursor keys, and function keys being adapted for specific tasks. The main Conquest program uses windowing techniques by which the estimator may easily view and select the data required. Within each window it is possible to page up and down the list of data and then highlight and select the required data item. Alternatively, the user may use 'Find' commands to locate the appropriate data item automatically. The user interface within the system provides the estimator with alternative methods of operation that allow the estimator to select the method best suited to his experience or individual requirements. The use of the windowing facilities with the mouse unit remove the need for complex coding of resources and work groups, so making the system more attractive to users and quicker to operate.

The facilities within the Conquest system

The Conquest system provides the following facilities to assist the estimator with the preparation of the direct cost estimate and the production of a tender:

- a library management system;
- alternative methods of data input;
- the ability to estimate the cost of the bill of quantities items;
- a full reporting system based upon a report generator;
- facilities to add an allowance for profits and on-costs; and
- system utilities.

(i)   The library management system

The Conquest system allows the estimator to store data in a data library for use in the pricing of bill items. Data within the library may be stored under a number of different classification formats including:

- the Civil Engineering Standard Method of Measurement (CESMM2) second edition;
- the Department of Transport Method of Measurement for Highway Works;
- the Standard Method of Measurement for Building Works, sixth edition (SMM6);
- the Standard Method of Measurement for Building Works, seventh edition (SMM7); or
- a classification unique to the estimator's own company or area of work.

Up to nine different libraries may be stored on the Conquest system and used by the estimator as a basis for pricing bill items. The estimator may switch between these libraries at any time. Data from different libraries may be combined within a single project estimate.

The library resources are individual resources or combinations of resources for which cost data are held on file. These resources are then combined into Work Groups for which production rates for each resource are entered. This enables typical build-ups to be produced which may be used as the basis for pricing bill items. The Work Groups are stored under classifications appropriate to the method of measurement selected.

Within the Conquest system, the materials and plant resources are stored in groups relating to the type of work for which they are to be used. When the system is used by the estimator to price bill items this results in only materials directly applicable to the classification of work being displayed. This makes the selection of required resources easier and quicker. If the estimator wishes to use a resource stored in a group relating to a different type of work the system allows these resources to be displayed via a 'window' on the screen and the appropriate resource selected.

Figs. 9.2 and 9.3 show sample screens from the Library Management suite of programs. These illustrate how the window facilities are used to select, combine and amend resource data.

(ii)   Method of data input

Data may be input into the Conquest system directly via the keyboard or indirectly via the Digital Building Systems OCR or digitizer systems. If either of these two linked systems are used to input bill data then a file of bill item data is created on the system and then transferred to the estimating system

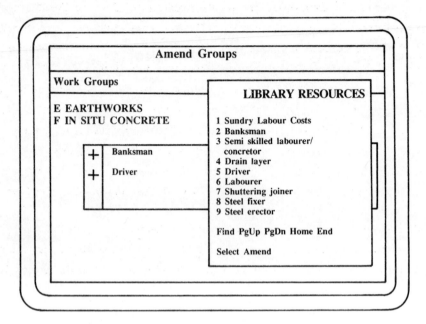

Fig. 9.2   The selection of library resources
(from the Conquest computer aided estimating system)

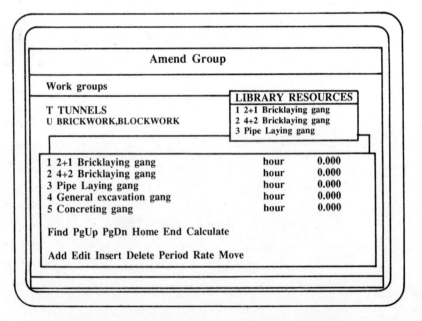

Fig.   9.3   The selection of library resources
(from the Conquest computer aided estimating system)

via a file transfer command. The task of entering bill data via the OCR may be delegated to the estimator's clerk. This frees the estimator to undertake other more important tasks such as building up all-in rates for labour and plant items or issuing quotation documents to sub-contractors. When the clerk has entered the bill data and made any minor corrections required, the estimator may proceed with the estimate. If the bill data are to be entered via the keyboard then this data entry is performed by the estimator within the Price Bill facility within the main estimating suite of programs. Data on each bill item are entered in turn and the estimator prices the bill item concerned before proceeding with the entry of data for the next bill item.

(iii)   Estimating the cost of bill of quantities items
The estimator is able to estimate the cost of the items within the bill of quantities in a number of different ways. These estimating methods include:
- unit rate estimating;
- spot or gash rate estimating;   and
- the use of sub-contractor quotations.

These methods may be used individually or in combination with each other.
Unit rate estimating is available in two forms:
(i)   by the use of previously recorded build-ups from the library management system;
(ii)   by the build up of a unit rate from first principles by nominating resources (labour, plant and materials) together with their associated output or usage rate.

Where the estimator wishes to use a spot rate to assemble an estimate for a bill item this may be achieved by supplying costs for the cost codes of labour, plant and materials without specifying the individual resources which will undertake the work. Alternatively the estimator may allocate selected sub-contractor quotations to bill items or to a group of bill items.

The usual method of system operation is as follows. Assuming that the estimator is inputting bill item data via the keyboard the appropriate command is selected and the project on which it is required to enter data nominated. The estimate for the items is undertaken by working within the appropriate method of measurement classification and using library data as the basis of each item build-up.

The estimator's build-up screen is displayed in the format of a page of a bill of quantities. The bill item description is related to the item reference code and the build-up rate, where available, taken from the data library. The rate is divided into the standard cost categories plus any additional categories previously nominated by the estimator. At any time the estimator may use the build-up command to display the resources used to price the item, together with their usage rates and cost rates. An example of an item display is shown in Fig. 9.4.

E	EARTHWORK				
	EARTHWORK GENERAL EXCAVATION				
	Topsoil				
E411.0	Maximum depth not exceeding 0.25m.				
	Labour 0.284	Plant 0.735	Material 0.000	Nom.sub 0.000	Sums 0.000
				1.019	

Fig. 9.4 An example of a bill item display
(from the Conquest computer aided estimating system)

The estimator has the ability to review fully the item build-up and make any amendments necessary to meet his requirements. This includes the changing of output and cost rates to the existing resources or the addition of new resources to augment the build-up. An indication of the flexibility of the system may be obtained by reviewing the list of ring menu and submenu options shown in Table 9.3.

If the estimator so desires the cost of the item may be estimated from first principles by the combination of individual resources or resource groups. If appropriate, sub-contractor rates or spot rates are entered.

Table 9.3   Ring menu and submenu options

(from the   Conquest computer aided estimating system)

OPTION	FACILITY
EDIT	Allows you to edit a line.
ADD	Allows you to add a new title or new block of data, as application to the screen.
INSERT	Allows you to insert a new line above the one highlighted.
DELETE	Allows you to delete the line highlighted.
EXIT	Allows you to exit from the current screen.
MORE	Gives further options.
MOVE	Allows you to move the highlighted line to a new position.
NEXT	Goes to the next level of data.
PREVIOUS	Goes to previous level of data.
CONSTANTS	Allows you to select specific resources.
GROUP	Allows you to create a new block of data or view or attach an existing block.
UNIT	Allows you to define or change the unit of measure for the highlighted line and data attached to it.
DATA TYPE	Allows you to define the highlighted line as an underlined title, a title not underlined, part of the item description or a non printed line.
EDIT TITLE	Allows you to edit the title of the group of data you are currently in.

All of the amendments required may be achieved by use of the mouse. This avoids any need for typing or the use of resource codes, and makes the changes very quick to execute. The use of the windowing system makes the selection of the required resources or outputs simple. The system anticipates the user's requirements when certain operations are selected. These 'intelligent actions' speed the execution of changes and remove the frustration of some systems that require the user to follow pre-determined paths through the system to make amendments. A 'Clone' facility enables the estimator to make additions to the item build-up using existing pricing information from items within the project under review or other projects for which data are held on the system. As changes are made there is an automatic re-calculation of the bill item rates and the bill summaries. Should the estimator need assistance in the operation of the system the 'Help' facility operated by a function key on the keyboard provides information relevant to the point in the system reached by the estimator. When the estimator is satisfied with the build-up of the bill item, the data are filed and the pricing of another bill item is commenced.

The estimator may review the estimate that has been produced in two ways. The 'View' facility lists the bill items in summary form. The estimator then selects the required item and views the item build-up. The system may also be operated in 'browse' mode. The estimator simply moves backwards or forwards through the bill items and reviews the relevant item data.

Conquest provides the facilities to locate and print out lists of sub-contract items identified by each type of sub-contractor. These lists of bill items may then be forwarded to different sub-contractors to obtain quotations for the work involved. Similarly lists of labour, plant and materials items required for the project may be produced. These listings enable other staff within the contractor's organization to collect and collate the data required by the estimator for the completion of the estimate.

The facilities for estimating the cost of bill items provided by the Conquest system and the user interfaces adopted represent considerable progress in computer aided estimating compared with the early command driven estimating systems that relied upon formal pre-defined procedures of system operation. The system is simple to operate. At any point the estimator may use either the keyboard or mouse to select the data or function required. Movement around the system is quickly and easily achieved allowing the estimator the full flexibility required when building up the direct cost of the items within the bill. At any time item costs may be adjusted. The cost of the project is then automatically recalculated.

(iv)   The reporting system

The Conquest system allows the user to print out information on individual bills of quantities, job resources and library information. Fig. 9.5 shows the screen format that the estimator uses to print out information from individual bills of quantities. Conquest includes a full report generator that allows the estimator to select the type of data, the level of detail included and the destination of the output: file, screen or printer.   Report formats may be stored within the system and used at a future date. This means that once the estimator has decided upon the format of the reports required for the tender, this information may be stored within the system and called upon whenever an updated report is required.   Figs. 9.6 and 9.7 show examples of printouts from the Conquest system.

Conquest includes a powerful 'search and select' facility which enables the estimator to select and report on items by item reference, the resources used, the quantity of work or the cost of the item.   For example, it is easy to produce a report on all bill items costing more than £2000 or all bill items priced by sub-contractor rates.

In the past, users of computer aided estimating systems were restricted to specific reports which contained information in pre-defined formats. Conquest uses the latest reporting techniques to provide the estimator with a complete range of information in the format required.

(v)   Facilities to add an allowance for profits and on-costs

Having produced a direct cost estimate for the work within the project, the estimator will wish to add additional monies to allow for profits and on-costs. The Conquest system allows the estimator to do this by the addition or subtraction of monies to:

- individual resource rates;
- individual bill item totals;
- individual bill sections;   and
- the overall direct cost of the project.

Alternatively the direct cost for the project may be amended by the addition or subtraction of percentages of bill section totals or the total cost of the project. In this way an estimate may be quickly and easily turned into

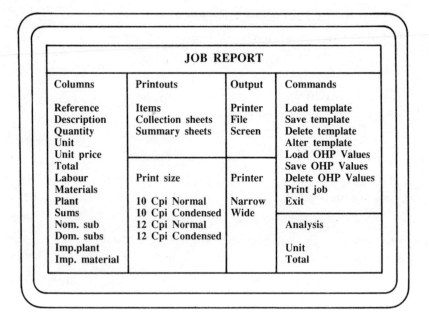

JOB REPORT			
Columns	**Printouts**	**Output**	**Commands**
Reference	Items	Printer	Load template
Description	Collection sheets	File	Save template
Quantity	Summary sheets	Screen	Delete template
Unit			Alter template
Unit price			Load OHP Values
Total			Save OHP Values
Labour	Print size	Printer	Delete OHP Values
Materials			Print job
Plant	10 Cpi Normal	Narrow	Exit
Sums	10 Cpi Condensed	Wide	
Nom. sub	12 Cpi Normal		Analysis
Dom. subs	12 Cpi Condensed		
Imp.plant			Unit
Imp. material			Total

Fig. 9.5   The screen format for reports
(from the Conquest computer aided estimating system)

a tender.   These additions may be made on a 'what if' basis and the full implications of these allowances for profits and on-costs examined.

(vi)   System utilities

Within the Conquest system there is a suite of system utilities designed to enable the user to operate the system efficiently and amend certain aspects of the system to meet the company's own requirements.   Some examples of the utilities provided are:

- the selection of display colours for the system screen displays;
- OCR column order;
- the number of decimal places displayed for the currency;   and
- system security and back-up facilities.

The user may set the system to operate with the screen displays, titles, menus and windows being viewable in a range of colours.   Default colours are set at the time of installation, but if the user prefers alternative combinations these may be selected and the system amended to display these until alternatives are desired.   If the user does not have a colour monitor monochrome options are available.

Pumping Station Siteworks

	D: Demolition and Site Clearance				Total
	Trees				
D210	Clearance of trees, girth; 500 mm - 1 m	38	nr	22.12	840.56
	E: Earthworks				
	General excavation				
E411	Topsoil maximum depth 0.1 m; from permanent site of pumping station for areas to be gravelled	72	m³	1.46	105.12
E412.1	Ditto 0.25 m; for areas of path	5	m³	1.46	7.30
E412.2	Ditto 0.33 m; for areas of concrete road	46.5	m³	1.54	71.61
E412.3	Ditto 0.33 m; for areas of block paving	370.3	m³	1.54	570.26
	Excavation ancillaries				
E521.1	Preparation of excavated surfaces topsoil	713	m²	1.21	862.73
E522.1	Ditto subsoil	1305	m²	1.21	1579.05
E531.1	Disposal of excavated material topsoil; to storage area at amenity bank	465	m³	2.85	1325.25
E541.1	Double handling of excavated material topsoil	32	m³	2.63	84.16
	In situ concrete				
	Provision of concrete – cement to BS4027				
F247.1	Designed mix Grade C20	40	m³	51.59	2063.60
F277.1	Designed mix Grade C35	104.5	m³	52.64	5500.88
F277.2	Designed mix Grade C35 minimum cement content 360 kg/m³	46	m³	53.98	2483.08
	Page total: 3/1				15493.60

Fig. 9.6   Sample report 1 from the   Conquest computer aided estimating system

Estimate for Bridgetown Sewage Works Phase 2

<u>**Pumping Station**</u>

<u>**Job: Sub-Structure**</u>	**Job Resources**	
	Unit	**Quantity**
<u>**Common Material**</u>		
Ordinary portland cement	tonne	7
Crushed limestone aggregate 40-5 max	tonne	29
Sand Zone 2/3. BS882	tonne	17
Sand/Lime motor	m³	16
<u>**In Situ Concrete/Large Precast Concete**</u>		
Ordinary prescribed concrete Grade 7.5	m³	22
Square mesh fabric – A142	m²	100
Formwork carcassing Timber (4 uses)	m³	5
Formwork plywood sheet (4 uses)	m²	84
12 mm mild steel dowel 60 mm long	nr	78
12 mm flex cell 150 mm wide	m	46
<u>**Masonry**</u>		
Sundry masonry material	£	150
Common bricks	1000	15
Galv. butterfly wallties	100	4
Facing bricks	1000	29
<u>**Reinforcement**</u>		
High Yield steel bars nominal size 12 mm	tonne	10.3
High Yield steel bars nominal size 16 mm	tonne	18.7
High Yield steel bars nominal size 20 mm	tonne	26.2
High Yield steel bars nominal size 25 mm	tonne	28.4
High Yield steel fabric type A393	m²	160
High Yield steel fabric type B196	m²	84

Fig. 9.7 Sample report 2 from the Conquest computer aided estimating system

Some bills of quantities documents are prepared with the quantity column printed before the unit column.   Others are printed with this information in the reverse order.   The Conquest system includes a utility to enable the user to inform the OCR of the format of the bill of quantities that is about to be read.

In   some circumstances the estimator may require the display and printout of resource costs and bill item totals to more than two decimal places. A utility command within the system allows the user to switch between two and three decimal places on displays and printouts.

The Conquest system recognises the importance of the estimator's data and provides comprehensive system security, including up to nine levels of password protection.   The back-up utility enables the user to copy data to and from floppy disks to ensure that a full back-up of library and project data may be held outside the system.

A summary of the Conquest system

Computer aided estimating systems are dependent on the storage, retrieval and manipulation of large quantities of data.   The Conquest system is based upon a relational database.   This offers the advantage of efficient management and manipulation of large data sets.   The developers have produced a system that is flexible and free from artificial constraints on data.   Although the system is based on a data library for use in the build-up of direct item costs, there are several alternative estimating methods available to the estimator. When the cost of the bill items has been established the estimator may review the item build-ups and make any necessary adjustments.   Last minute changes in resource prices may be easily accommodated.   This has resulted in a system that is easily used by estimators and suitable for integrating with other software.

LINKS BETWEEN ESTIMATING AND OTHER MANAGEMENT FUNCTIONS

The estimating department is the first department in a contractor's organization to have contact with a project and the contract documents describing the project.   The estimators not only produce an estimate of the cost for use in a tender but also in doing so they assemble and organise a considerable amount of pertinent data and information relating to the project.

These data, gleaned from contract documents, promoters' representatives, suppliers, sub-contractors, and also provided by the estimators, are used primarily for the production of the estimate. These data are also of value to subsequent job control activities. However, the use of these assembled data in subsequent job control activities is limited. The reason for this is that in most manually produced estimates the assembled data are difficult to access or interpret. These difficulties have led in many cases to a clear and unnecessary division between estimating and subsequent job control activities such as planning, cost control, measurement and valuation. In companies where these subsequent job control activities are undertaken, at least partially, with the aid of a computer, the task exists of entering a considerable amount of data for each contract awarded.

The existence of computer aided estimating offers the facility of retention of detailed, meaningful and accessible records of the estimator's calculations and assembled data. Thus subsequent job control activities will benefit from establishing computer aided estimating. These benefits are becoming a main consideration in companies considering such systems. It is important, therefore, that the estimating system installed is capable of providing the downstream functions with the data required.

Fig. 9.8 shows the logical links between estimating, planning, cash flow, cost control and measurement.

Providing these systems are capable of accessing the data files of the estimating system and reading information on the files of contract data, the estimator's data may be used in these 'downstream' management functions. An outline of each of these functions is now given.

Measurement, valuation and price adjustment

These three aspects require details of the bill, item by item. These details already exist in the files of contract details and are in computer compatible form. In manual estimating systems they have to be entered into a computer before computer systems can be used to assist in these tasks. In price adjustment, for example, the volume of data required to be entered virtually makes it uneconomical to use computers, unless the data have already been entered for other reasons. The only information that is needed to supplement that stored in the file of contract details is the work category used in the formula price document.

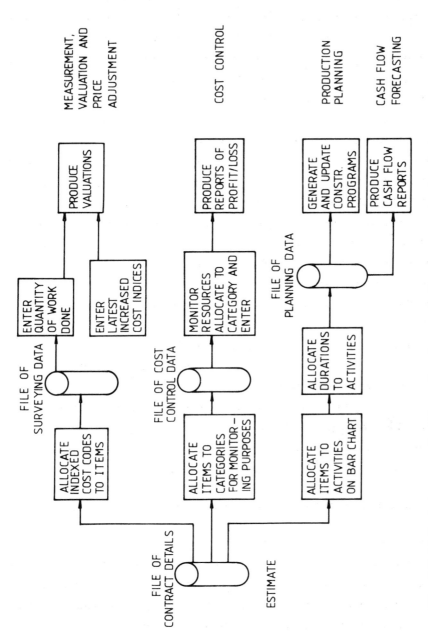

Fig. 9.8 Links between estimating, planning, cash flow, cost control and measurement

Cost control

The task commonly known as cost control comes in several forms.   The simplest is cost monitoring whereby the actual labour, plant and material costs are collected and compared to the budgeted costs in the form of a variance analysis.   It is in the establishment of the budgeted costs that the estimator's records are invaluable.   If they are not accessed from the files of contract details, they have to be re-entered into the cost control system.   If they are lost altogether, they have to be re-established.   Thus computer based cost control systems stand to benefit from computer aided estimating.

Production planning

Numerous stand alone planning packages exist, most of which are based on network planning.   The link between planning and the estimator's data contained in the files of contract details is achieved by allocating each item in the bill of quantities to an activity in the network (or bar chart).   Each activity then is a collection of items with their constituent resources.   The resources in each activity are then aggregated to form the resource requirements for each complete activity.   This enables the network to be manipulated in order to achieve a suitable flow of resources.

The aggregation of resources on an activity by activity basis also enables the planner to assess the activity duration produced by the estimator. This is then the subject of further discussions between the estimator and planner.

Thus the link between the estimator's data and the construction programme provides a continuity of data which, until the advent of computer aided estimating, was extremely time consuming and tedious to achieve.

Cash flow forecasting

Cash flow forecasting needs to be divided into two aspects.   One is the cash flow forecast for an individual contract and the other is the cash flow forecast involving the amalgam of many contracts and other activities.

Cash flow forecasting for an individual contract

The data contained in the files of contract details include those required for a cash flow forecast, except that each bill item needs to be allocated to activities on a bar chart or network in order to determine the time of each item of work.

If this is done and in addition details of credit delays and contract payment conditions are given, it is then possible to devise computer programs to produce cash flow information.

Some planning packages also produce cash flow forecasts and, if such a package is used, obtaining a contract's cash flow is a simple by-product of planning work.   Thus, the previous comments applied to planning equally apply to cash flow forecasting.

The common use of spreadsheet packages for producing cash flows demands the need for the estimating system to produce files in a format that may be accessed from within the spreadsheet to allow the project data to be read and utilised within the spreadsheet calculations.

Cash flow forecasting for company regional or divisional offices

This can be regarded as a combination of the individual cash flows from individual projects.   More usually this is derived from accounting data held elsewhere within the system.

Computer integrated construction

One example of an integrated computer based management system is marketed under the product name 'Computer Integrated Construction'. Computer Integrated Construction (CIC) has been developed by Digital Building Systems supported by a major hardware manufacturer.   The system operates under the UNIX operating system and adheres to 'open systems' guidelines.   The estimator's data from the Conquest system are used as the basis for an integrated construction management system.   The CIC system, introduced in 1989, selected stand alone industry recognised applications and combined them into an integrated system which permitted different functions to access, store and distribute information.   The modules include:

- Estimating and Valuation;
- Project Management and Resourcing;
- Plant Management;
- Financial Management;   and
- Office Automation.

Within each of these applications, access to individual modules is controlled by a security system based upon the user's name and password. Links are made between each module and specific commands allow the transfer of data from one module to another. Each integration transfer will check that the latest data are to be transferred and that the data are complete. Error messages warn users where data are not current and file locking facilities ensure that other users of the system are not using the files whilst they are being updated. Typical links within the CIC system from the estimating module include:

- cost estimates to the job contract ledger;
- activity and resource information to project management;
- material requirements to purchase orders;
- project plant requirements to the plant management module;   and
- estimating information to the valuations system.

The project cost analysis information produced by the Conquest system is passed to the job costing module and used as a basis for budget costs. The data are transferred under user control when the costing analysis has been completed. This will automatically set the budgets from each cost heading transferred.

Data on construction activities and individual resources from each estimate are transferred under user control when the bill pricing is completed. The project manager groups the activities as required and adds the respective dependencies. The bulk transfer of these data then sets up the basic data within the files of the project management system.

The user may transfer the total material requirements for the project from the estimate to the purchase ordering module for the purpose of creating bulk or call off orders. 'Skeleton' purchase orders are created from the details passed and the user supplies additional information to create the completed purchase order document.

Details of the plant requirements on the project are transferred from the estimating system to the plant management module to give advance warning of the plant needed for the project. These data are accessed by the plant manager either on a screen or on a printed copy of the report. From this information the manager is able to forecast future plant movements and arrange for hired plant to make up any shortfall in availability.

The estimating system is fully integrated with the valuations module.

All bill items are accessible when the valuation is being entered, and are displayed on the screen to assist in the data entry process. This necessitates only the valuation data to be entered by the user to produce the monthly measurement.

Within the CIC system data integrity is maintained by the Integration Support System that regularly checks the data in the different modules to ensure that the values held are consistent. The system includes full reporting facilities which enable data to be retrieved from different systems and included with a single report. These reporting facilities allow management to access, manipulate and utilise the information summarising the financial status of the project at any level of detail required.

THE IMPLEMENTATION OF COMPUTER AIDED ESTIMATING SYSTEMS

The installation of a computer aided estimating system in a contractor's office where none previously existed and the transfer from manual to computer aided estimating is not a simple task. Many estimating personnel are inexperienced in the use of computers, and the computing experience that exists in construction companies tends to reside within accounts departments and computer service departments. Thus the introduction of computer aided estimating to estimators also involves the introduction of computers. The aspects that must be taken account of in the development of implementation procedures are as follows:

(i)   The implementation of computer aided estimating is a process involving several stages and is not achieved in one step.

(ii)   There are at least three types of personnel involved in running a computer aided estimating system: these are the senior manager or the chief estimator, the estimator and the estimator's clerk. Each type of user requires different training and support.

(iii)   There are many different forms of training and support such as instruction manuals, within system aids, formal training, provision of local experts and access to computer advisory personnel. No one form of user support is appropriate to all types of user in all phases of implementation and, therefore, the appropriate forms of support must be provided to each user at the appropriate times.

(iv)   The estimating personnel with little or no computer experience have a number of legitimate fears or concerns relating to the use of computers. These include dealing with the complexity of the estimating process, retaining the ability to exercise judgement, coping with errors, meeting deadlines, security problems and re-training difficulties.

This section considers the following aspects of the implementation of computer aided estimating systems:

- the type of user;
- the phases of implementation;
- support requirements; and
- estimators' fears of computer aided estimating.

Types of user

The estimating and tendering process requires the estimator to collect data from a number of sources and produce a direct cost estimate for the works within the project. This estimate is, following the review by senior management, turned into a tender by the addition of allowances to cover overheads and profits. Most systems do not assume that the estimator will be responsible for all the tasks to be undertaken on the computer system. Experience has shown that there are typically three types of user of computer aided estimating systems:

- the estimator's clerk who maintains files of data and operates peripheral hardware such as the OCR;
- the estimator who inputs the material prices, selects resources and usage rates for the bill items and prepares reports for the tender adjudication meeting; and
- the contractor's senior manager who inspects item build-ups, makes amendments to the direct cost build-ups and adds mark-up factors for the contract.

When a system is implemented there is invariably an imbalance between the knowledge and skills of the people who will be operating the system and the complexity of the system to be introduced. The degree of imbalance determines the support and assistance required by each of the different types of user.

Several different types of support exist. The main ones for computer aided estimating systems are:

- a comprehensive instruction manual;

- within system aids;
- formal training instruction;
- provision of a local expert;   and
- access to computer advisory personnel.

None of these methods of user support alone is satisfactory in meeting the needs of each type of user at the various stages of their development with a system.

Phases of implementation

The different phases of implementation which delineate different stages of development and so require different training and support are:

- the pre-implementation phase;
- the implementation phase;
- the operational phase;   and
- the evolutionary phase.

The following sections describe these phases and the different user support requirements for each type of user in each phase.

Pre-implementation phase

The pre-implementation phase is when a company first seriously considers using a new system. The success of implementing and using an estimating system is dependent on the estimators and clerks who will be responsible for inputting data, performing calculations and preparing reports. The standard of information supplied to others within the company is also dependent on the work of these estimators and clerks. It should be obvious that these 'key' users have to be consulted and involved in the decision to implement.

In the implementations of computer aided estimating systems studied by research staff at Loughborough University of Technology it was seen that at this stage the estimators and clerks raised information queries on issues such as the range of facilities available in the system and the operational requirements placed on them, and the degree of control they would have over the system. In some cases where a lack of understanding existed, a lack of interest and potential aversion emerged. Discussions on the users' job security and job satisfaction are also required.

In this phase all three types of personnel, the clerk, the estimator and the senior manager, often require demonstrations and test runs to satisfy themselves that the computer system can perform the tasks of estimating. This invariably involves preparing a test estimate for a contract. Such trials have the effect of proving that the software is satisfactory. Another method of achieving this is to visit existing installations and to have discussions with established users. This may only be partly satisfactory since systems may sometimes be refined and updated. If the selection of hardware is seen as a separate issue then the company and the personnel involved have the additional problem of surveying the hardware available.

The pre-implementation phase can be protracted. The demands made by the construction companies on the suppliers of software and hardware can be extensive.

Implementation phase

When a system is installed, the estimators and clerks will be preoccupied with learning the operating procedures and re-adjusting their normal duties to accommodate the new system. During this phase, the earlier fears of these users are either dispelled or confirmed. This seems to depend not only on the quality of the system but also on the support provided to aid assimilation of the new skills. There is a natural tendency for people to resist change and hence the level of support provided at this early stage is important so that the new system can be quickly and efficiently learned. The speed and ease of learning has emerged as a major factor in determining the attitude of the users towards the new system.

It is during this phase that other organisational and developmental difficulties are faced. For example, most systems offer the company the opportunity of storing 'performance data' for use by the estimators. In companies where 'company manuals' already exist, the inclusion of these data in a computer system is no more than the substantial but mechanical task of translating the existing data into a format suitable for entry into the computer system and the task of entering the data. In other companies where such manuals do not exist the task of capturing and collating the performance data which exist within the company in the form of estimators' individual files or records and in estimators' memories is a larger task. Some companies take 'test' data from other sources to 'get them started'. This has proved to be counter-productive since estimators often argue about the quality of the data, which they eventually replace anyway.

The major pressure during this stage comes from maintaining the normal output of the estimating department while trying to accommodate the new system. This normally requires at least one man to be relieved of his normal duties. This can be more easily achieved in the larger companies than in the smaller ones. It is a time when most staff have to bear extra pressure.

Operational phase

The operational phase is reached when a system is functioning within the organisation to the extent of meeting all the primary objectives such as producing estimates and tenders. The presumption is often made by commercial software houses that user support requirements are now minimal. However, it is clear that other support requirements become evident. Examples are non-routine events such as breakdowns or errors that may produce situations that were not catered for in previous training. In addition, estimators sometimes find that the routine functions of the system fail to meet all their needs now that their knowledge and skill have been extended. Thus greater understanding of a system leads to consideration of refinements or extensions to the facilities available.

Evolutionary phase

The evolutionary phase is an extension of the operational phase and is when the almost inevitable system modifications are defined and consequently a need for further advice and technical skill arises. The evolutionary phase renews the cycle of development. User support is an on-going requirement, not solely an action necessary at the initial stage of the introduction of a system.

For companies who choose limited or restricted systems in the first place the evolutionary phase may well represent buying a completely new system. It is clear that the estimating personnel's ability to use computer systems is frequently underestimated and this leads to purchase of the cheaper, limited micro computer systems which cannot be expanded. One major limitation of the smaller computer systems is that they only retain limited records of the estimator's calculations. This prevents the estimating system passing data to other company functions such as cost control, planning or cash flow. For a company that is new to computing such advantages may not be apparent at the stage of selecting the first estimating system. Once some experience and competence in the use of computer systems has been gained these advantages emerge and so the company outgrows its computer facilities. This results in the purchase of larger and more flexible systems and begins the process of implementation again, albeit from a more informed stance.

User support requirements

Most computer systems are designed to meet the needs of only one type of user. The estimating and tendering process within a contractor's organisation requires a system that is used by three types of user.

The various forms of user support that exist, which are relevant to the use of computers in estimating, are:

- instruction manuals;
- within system aids;
- formal training instruction;
- 'local expert';   and
- computer advisory personnel.

Instruction manuals

A comprehensive reference manual is required for computer aided estimating systems to state formally the facilities that are available and the manner in which the system should be used to perform the tasks of the estimator's clerk, the estimator and senior management. There are disadvantages with this type of support or aid relating to the need to consult a document which must be comprehensive and, therefore, weighty. Instruction manuals are not nowadays used to guide the user through the input as they were in 'batch' systems of old, because the 'interactive systems' provide 'prompts' or instructions on the VDU screen. Nevertheless, the manual is required as a fall-back with which to deal with problems.

Within system aids

Most systems provide some form of within system aids to assist the user.

Different types of aids include: menu commands, error messages and a comprehensive 'Help' facility.   The availability, consistency and comprehensiveness of these messages distinguish a good system from an inferior one.

Formal training instruction

The training requirements of the different types of user, clerk; estimator and senior management, are different.   Most suppliers of systems provide formal training instruction.   The content of these training programmes should reflect the type of user.   Each type of user should have a detailed list of operations within the system so that all concerned can be confident that the users, after training, are capable of operating the system.   For example:

> An estimator can be considered proficient in the operation of the system when he can:
> -   operate the hardware without assistance in its normal mode of operation;
> -   locate the contract for which he was required to produce an estimate;
> -   adjust all the prices of the resources for the contract;
> -   understand the structure of the system to enable any command to be located;
> -   understand the facilities offered by each command and perform the tasks with each command;   and
> -   produce an estimate unaided.

A comprehensive training scheme should include:
> -   lectures and demonstrations with appropriate audio visual aids;
> -   informal group discussions;
> -   guided tuition on the operation of the system;   and
> -   structured exercises undertaken by each trainee.

Local expert

The introduction of any computer system leads to the evolution of a 'local expert'.   It is therefore logical that, rather than let this disrupt the normal work pattern of the 'expert', the position is formally recognised within the company.   This provides a contact point between the company and the suppliers of the system and enables specific problems to be solved as well as providing a point of reference for all other users within the company.

Computer advisory personnel

After implementation of estimating systems the estimators still need access to computer advisory personnel. As the estimators' understanding of systems grows, their needs evolve and eventually they will require amendments to the system to incorporate other aspects of their tasks which could be performed within the system. It is important that a continuing liaison is maintained between the estimators and computer advisory personnel, whether they are within the company or from the supplier of the system, in order that the system can be extended when required and so used to its full capacity.

Summary of user support

Fig. 9.9 summarises the user support requirements for each type of user in each stage of implementation.

Estimators' fears in relation to computers

A survey of estimators prior to the implementation of computer aided estimating systems indicated six major fears with respect to the introduction of computers.   Each of these fears is described below, followed by the resolution of the perceived difficulties.

Complexity of the estimating process and data availability

The process of estimating and tendering has been seen as beginning with the invitation to tender and ending with a tender submission.   During the preparation of the cost estimate and then the tender, the estimating department has exchanges of information with planning departments, purchasing departments and company directors, and materials suppliers externally.   The estimators refer to their historical records, the company manuals and other supporting data.   The difficulties foreseen by estimators were external information not being available on time, or information arriving

PHASE OF IMPLEMENTATION

USER TYPE	Pre-implementation	Implementation	Operational	Evolutionary
Estimator's Clerk	Explanations Demonstrations Reassurance	Formal Training Within System Aids	Within System Aids Local Expert	Local Expert
Estimator	Explanations Demonstrations Reassurance Trials	Formal Training Within System Aids	Within System Aids Local Expert Reference Manual	Local Expert Computer Advisory Personnel
Senior Management	Explanations Demonstrations Reassurance Trials	Formal Training Within System Aids	Within System Aids Local Expert Reference Manual	Local Expert Computer Advisory Personnel

Fig. 9.9 A summary of user support requirements for each user in each stage of implementation

in different sequences or changes to the proposed method of executing the work.

Resolution

Highly flexible systems are required that enable data to be entered and amended at any stage. Information arriving in different sequences or relating to different methods of construction can then be easily incorporated without disrupting the estimator's work pattern or restructuring his tasks within the estimating and tendering process.

Computers and estimators' judgement

Estimating was seen by the estimators as being like some aspects of design where the process is a mixture of calculation and judgement. The estimators were very concerned about their opportunity to exercise this judgement.

Resolution

These fears have been dispelled by use of flexible systems with estimators operating VDU's using comprehensive editing facilities to manipulate stored data.

Computers and errors

Estimators feared the following:

(a)   errors in data which are used over and over again;   and

(b)   computer programs which on receiving input data from the estimator would perform several calculations and give the estimator some results. The estimator would not know precisely what had been done to produce the results, how the answer was arrived at, and even if it was correct.

Resolution

The use of computers for routine calculations prevents manual calculation errors. In the system comprehensive displays and printout facilities should enable the estimator to check the resources, output or usage rates and calculations within each item build-up. Where any code is used to retrieve data the nature of the item should be displayed with descriptions in 'intelligible' English.

Computers and tender deadlines

The estimators surveyed were keen to establish that the deadline for submitting tenders is very precise and that the estimating process must take place within the time available. There could be no delays that would cause the tender to be late. Thus the users of computer aided systems insist that if any failure occurs, particularly at a late stage in a tender's preparation, the estimators have a means of recovery and sufficient information to complete the tender.

Resolution

Correct back-up facilities will ensure that contract data are copied and held secure against equipment failure. Prudent use of the reporting facilities as the tender deadline approaches ensures the latest details of the tender are always available as hard-copy printout should any untoward occurrence affect the performance of the system. The introduction of networked micro-computer systems to run the software ensures that there is duplicate hardware available on which to produce the final tender reports.

Computers and security

The estimating process was seen by the estimators to have security problems, particularly with respect to the management of data files (discs, etc.). The security problem was seen as being particularly acute because the calculations related to tenders not yet submitted, and estimators were very aware of the value of the information to competitors.

Resolution

The most widespread and acceptable resolution is to provide estimators with their own computers installed in their offices and dedicated to the task of estimating. Fears on security are overcome by ensuring that all hardware and discs are physically in the estimator's own control.

Estimators and re-training

The estimators realised that the introduction of computers would require them to be trained in the use of equipment and systems. The extent of this training was a source of concern.

Resolution

Although estimators do not need to be trained as programmers they need to use the input keyboard and to appreciate the ability of computers to handle stored or filed data and to create and copy other files. This is not in itself difficult but does require some training. The production of a very flexible system operated by easily understandable commands, together with a structured training programme, means that estimators can be confident in using the system within a short time period. Nevertheless re-training is required.

OUTCOME

The successful implementation of computer aided estimating systems requires that appropriate consultation, support and training be provided to the three types of user, clerk, estimator and senior manager, at each of the four implementation stages, viz. pre-implementation, implementation, operational and evolutionary. Many systems are installed with less than adequate attention being paid to the implementation procedures and training required. This is caused, in the main, by low profit margins on software being sold to construction companies and the companies themselves not recognising the degree of training and support required until it is too late and a less than satisfactory implementation experience has been lived through. This is one major reason for computer systems failing to achieve the intended performance levels in a reasonable time.

THE FUTURE FOR COMPUTER AIDED ESTIMATING

During the last decade computer aided estimating has made a significant impact on the construction industry.   With the continuing development of information technology the acceptance and growth of computer aided estimating can be expected.   The following aspects of information technology are expected to play a significant role in future systems:

- improved user interfaces;
- user configurable systems;   and
- the use of expert or knowledge based systems.

Improved user interfaces

Estimators within the building and civil engineering industries have indicated their reluctance to adopt systems which they consider to be too cumbersome to use.   As software developers have become more aware of estimators' requirements they have developed improved user interfaces using the latest software techniques to ensure that their products meet the demands of the market place.   This aspect of systems development is expected to continue in the future with both conventional forms of interface and more futuristic forms of interface such as touch-screen and voice input.

User configurable systems

Although the estimating and tendering process is fundamentally the same in any contractor's organisation, different companies have different requirements from computer aided estimating systems.   Estimators involved with the development of bespoke systems have the opportunity of assisting in the design of their system, thereby ensuring that it will meet their exact needs. Estimators involved with the purchase of applications packages are faced with the need to adapt their methods of working to meet the requirements of the system.

There are those that consider this situation indicates that the current generation of applications packages represents a failure on the part of information technology to supply satisfactory systems to estimators (62).   They stress that the estimator should dictate to the system the method of working and not the other way round, as is the case at present.

Recent advances in computer science theory have adopted the use of 'fourth generation' languages and relational databases to construct 'configurable declarative' systems (67). This means that the systems have a built in flexibility and may be changed or configured to meet individual estimators' requirements in the functionality and operational features available within the system. No such systems are as yet commercially available but prototypes have been designed and these show the future direction for systems that will be easily amended to meet estimators' requirements. These types of system will also permit the development of systems that allow estimating, planning and others to be built together.

The use of expert or knowledge based systems

A recent study of the use of computer aided estimating systems by civil engineering estimators (73) has confirmed the area of estimating and tendering for civil engineering works as suitable for the use of knowledge based expert systems. This survey of the members of the Federation of Civil Engineering Contractors showed that existing computer aided estimating systems, although assisting the estimator with tasks dominated by arithmetical calculations, failed to accommodate uncertainty, or to work satisfactorily with incomplete data or allow estimators the full facilities to exercise their expertise. The survey showed that within the estimating and tendering process the most important tasks all relied upon a high level of estimator expertise. To date this issue has been almost totally ignored by developers of systems for civil engineering estimators. As the development of expert or knowledge based systems continues it is inevitable that such systems to assist the estimator will be produced.

The following tasks within the estimating and tendering process all represent suitable areas for system development:
- the decision to tender;
- materials enquiries;
- the selection of sub-contractors;
- evaluation of the conditions of contract; and
- the assessment of the estimate and the evaluation of adjustments and contribution.

Prototype knowledge based systems that have already been produced include those for the selection of earthmoving equipment (64) and the selection of cranage for construction sites (63).   Tah has indicated the possible use of expert systems for estimating drainage works (65).   It is current practice for estimators to select their labour and plant requirements for drainage works by consideration of the specifications for the pipe, the pipe diameter, the depth of trench required and the soil conditions.   These parameters determine the type of plant required and the output rates to be used within the estimate. This type of selection is particularly appropriate to the use of knowledge based systems.   Future computer aided estimating systems will not simply calculate item build-ups based on libraries of construction activities.   They will assist the estimator in the selection of the resources required and then use the appropriate output or usage rates to produce the initial estimate of the cost of the work.

10 References and Bibliography

The following is a list of relevant books and articles grouped into general headings. While some references are specific to building contracts it is considered that they contain some relevance to the civil engineering estimator.

A. GENERAL BOOKS

1. Harris, F. and McCaffer, R. (1989), **Modern Construction Management,** Third Edn: BSP Professional Books, Oxford.

2. Harris, F. and McCaffer, R. (1986), **Worked Examples in Construction Management,** Second Edn: BSP Professional Books, Oxford.

3. Harris, F. (1989), **Modern Construction Equipment and Methods,** Longman Scientific and Technical Publishing (UK), Wiley (USA).

4. Harris, F. and McCaffer, R. (1990), **Managing Construction Equipment,** MacMillan.

B. CONTRACT AND METHODS OF MEASUREMENT

5. The Institution of Civil Engineers (1985), **Civil Engineering Standard Method of Measurement CESMM2,** Second Edn: ICE/Federation of Civil Engineering Contractors.

6. Department of Transport (1987), **Method of Measurement for Highways Works,** Third Edn: HMSO, London.

7. Department of Transport (1988), **Notes for Guidance and Library of Standard**

Item Descriptions for Highway Works: HMSO, London.

8. The Institution of Civil Engineers (1986), **ICE Conditions of Contract,** Fifth Edn: ICE, London.

9. The Institution of Civil Engineers (1986), **Civil Engineering Procedure,** Fourth Edn: Thomas Telford, London.

10. Economic Development Committee for Civil Engineering (1971), **Report of Price Adjustment Formulae for Civil Engineering Contractors:** NEDO.

11. McCaffrey, R., McCaffrey, R. and McCaffrey, M. (1988), **CESMM2 in Practice,** BSP Professional Books, Oxford.

12. Reynolds, G.J. (1980), **Measurement of Civil Engineering Work:** Granada.

13. Uff, J. (1981), **Construction Law,** Third Edn: Sweet and Maxwell.

14. Abrahamson, M.W. (1974), **Engineering Law and the ICE Contracts,** Fourth Edn: Applied Science Publishers Ltd.

15. Barnes, N.M.L. (1971), **The design and use of experimental Bills of Quantities for civil engineering contracts:** PhD Thesis, UMIST.

16. The Royal Institution of Chartered Surveyors and National Federation of Building Trade Employers (1979), **Standard Method of Measurement for Building Works,** Sixth Edn.

17. Royal Institute of Chartered Surveyors (1988), **Standard Method of Measurement of Building Works SMM7.** Royal Institute of Chartered Surveyors and Building Employers Federation.

C. PLANNING

18. O'Brien, J. (1984), **CPM in Construction Management**, Third Edn: McGraw-Hill.

19. Cormican, D. (1985), **Construction Management : Planning and Finance:** Construction Press.

20. Adrian, J. (1981), **CM: The Construction Management Process**: Reston Publishing Co Inc.

21. Harris, R.B. (1978), **Precedence and Arrow Networking Techniques for Construction:** Wiley.

22. Laufer, A. and Tucker, R.L. (1987), **Is construction project planning really doing its job? A critical examination of focus, role and process**: Construction Management and Economics, 5, 243-266.

23. Laufer, A. and Tucker, R.L. (1988), **Competence and timing dilemma in construction planning**: Construction Management and Economics, 6, 339-355.

24. Willis, E.M. (1986), **Scheduling Construction Projects**, John Wiley.

25. Moder, J.J., Phillips, C.R. and Davis, E.W. (1983), **Project Management with CPM, PERT and Precedence Diagramming,** Third Edn: Van Nostrond Reinhold.

26. Neale, R.H. and Neale, D.E. (1990), **Construction Planning**: Thomas Telford, London.

D.   ESTIMATING, TENDERING AND RELATED SUBJECTS

27. Spence Geddes (1981), **Estimating for Building and Civil Engineering Works:** Newnes Butterworth.

28. Slattery, F.J. (1978), **Fundamentals of Estimating for Engineering Construction**: Australian Federation of Construction Contractors.

29. Thompson, P. (1981), **Organisation and Economics of Construction**: McGraw-Hill.

30. Civil Engineering Construction Conciliation Board for Great Britain (1981), **Working Rule Agreement**.

31. Estimating Practice Committee (1983), **Code of Estimating Practice**, Fifth Edn: Chartered Institute of Building, Ascot.

32. Abdel-Razek, R and McCaffer, R. (1987), **A change in the UK construction industry structure: Implications for estimating:** Construction Management and Economics, 5, 227-242.

33. Ashworth, A. *et al.* (1982), **Accuracy in Estimating**, Occasional Paper No. 27: Chartered Institute of Building.

34. Azzaro *et al.* (1987), **Contractor's Estimating Procedures: An overview:** RICS.

35. Bentley, J. (1987), **Construction Estimating and Tendering**: E & F.N. Spon, London.

36. Harrison, R.S. (1987), **Managing the Estimating Function**: CIOB, TIS No. 75.

37. Thomas Telford Ltd (1989), **Contractors file,** New Civil Engineer, February 1989.

38. Barret, N. (1989), **Sunrise sunset**, New Civil Engineer, February 22 1989: Thomas Telford Ltd.

39. Federation Internationale d'Ingenieurs Conseils (1977), **Conditions of Contract (International) for Works of Civil Engineering Construction,** Third Edn: FIDIC.

40. Clay, E. (1983), **Post-tender use of estimating data**, MSc Construction Management Course, Loughborough University of Technology.

41. Mudd, R.D. (1984), **Estimating and Tendering for Construction Work**: Butterworth.

42. Hardie, G.M. (1987), **Construction Estimating Techniques**: Prentice Hall.

43. Cullivan, D.E. (1981), **How much will it cost?**: Consulting Engineer, February.

44. Cox, V.L. (1982), **International Construction Marketing, Planning and Execution**: Construction Press.

45. Osumi, H. (1988), **An appreciation of an overseas project. Documentation, planning, estimating and tendering in a case study**: MSc Project Report, Construction Management Course, Loughborough University of Technology.

46. Jones, G.P. (1979), **A New Approach to the International Civil Engineering Contract:** Longman Group.

47. Sharpe, K. (1984), **Pre-tender information required for overseas contracts**: MSc Project Report, Construction Management Course, Loughborough University of Technology.

E.   CASH FLOW

48. Kaka, A.P. and Price, A.D.F. (1990), **Net cash flow models : are they reliable?** Accepted for publication in Construction Management and Economics.

49. Oliver, J.C. (1984), **Modelling cash flow projections using a standard micro computer spreadsheet program**: MSc Project Report, Construction Management Course, Loughborough University of Technology.

50. Kenley, R. (1986), **Construction project cash flow modelling**: PhD thesis, University of Mebourne.

51. Kenley, R. and Wilson, O. (1986), **A construction project cash flow model - an ideographic approach**: Construction Management and Economics, 4, 213-232.

52. Ng, L.K. (1983), **Cash flow forecasting using micro-computers**, MSc Project

Report, Construction Management Course, Loughborough University of Technology.

53. Allsop, P. (1982), **Cash flow and resources aggregation from estimating data**: MSc Project Report, Construction Management Course, Loughborough University of Technology.

54. Abernathy, E.S. (1990), **Managing a contractor's billings**:   Journal of Accountancy, February.

F.   ESTIMATING AND COMPUTERS

55. Bowman, S. (1981), **Tianareba: time analysis and resource balancing,** MSc Project Report, Construction Management Course, Loughborough University of Technology.

56. Sher, W. (1981), **Interactive estimating for building contractors,** MSc thesis, Loughborough University of Technology.

57. McCaffer, R. and Sher, W. (1981), **Computer aided estimating - an interactive approach:** Building Technology and Management.

58. Baldwin, A.N. (1982), **Computer aided estimating for civil engineering contractors:** PhD thesis, Loughborough University of Technology.

59. Baldwin, A.N. and McCaffer, R. (1984), **The implementation of computer aided estimating systems in construction companies**: Proc. Instn Civil Engrs, Part I (76), February, 1984, 237-248.

60. Baldwin, A.N. (1986), **Linking estimating and planning:** Construction Computing, Autumn 1986, 22-24.

61. Jabri, M.E. (1981), **Post tender use of estimator's data,** MSc Project Report, Construction Management Course, Loughborough University of Technology.

62. O'Brien, M. (1989), **A new approach to estimating systems**: Construction

Computing, No. 26, Summer 1989, 18-19.

63.   Cooper, C.N. (1987), **Cranes - a rule based assistant with graphics for construction planning engineers:** AI-Civil Comp. Proceedings of 3rd Int Conference on Structural and Civil Eng. Computing, Civil Comp Press, Edinburgh, 47-55.

64.   Alkass, S.T. (1988), **An expert system applied to earthmoving operations and equipment selection**: PhD thesis, Loughborough University of Technology.

65.   Tah, J.H.M. (1986), **Subroutines for estimating, valuations and cost control**, MSc Project Report, Construction Course, Loughborough University of Technology.

66.   Baldwin, A.N. and Oteifa, S.A. (1990), **The expert system approach to estimating and tendering for civil engineering works**: Unpublished paper, Department of Civil Engineering, Loughborough University of Technology.

67.   Pantouvakis, J.P. (1990), **Declarative-configurable estimating systems for the construction industry**: PhD thesis, Department of Computer Science, University of Nottingham.

68.   Morris, A. (1989), **So you want to buy an OCR?**: Construction Computing, No. 25, Spring 1989, Chartered Institute of Building.

69.   Morris, A. (1990), **OCR Revisited**: Construction Computing, No. 28, Winter 1989/90, Chartered Institute of Building.

70.   Wager, D. (1986), **First generation estimating**: Construction Computing, Spring 1986, Chartered Institute of Building.

71.   CICA and KPMG Peat Marwick McLintock (1987), **Building on IT (A survey of information technology trends and needs in the construction industry)**: Peat Marwick McLintock.

72.   CICA and KPMG Peat Marwick McLintock (1990), **Building on IT - For the

90's (A survey of information technology trends and needs in the construction industry): Peat Marwick McLintock.

73.   Oteifa, S. and Baldwin, A.N. (1991), **Estimators' tasks and computer aided estimating systems : A survey of FCEC member companies**. Construction Management and Economics, Volume 9 Number 3/4.

74.   Scott, G. (1990), **The selection and testing of application software packages:** Final Year Project Report, Department of Civil Engineering, Loughborough University of Technology.

75.   CICA and RIBA (1990), **Construction industry software selector 1990/91**; RIBA Services Ltd.

76.   Ndekugri, I. and McCaffer, R. (1986), **Valuations - An interactive system linked to estimating**: Construction Computing, July 1986.

77.   Allwood, R.J. (1989), **Techniques and Applications of Expert Systems in the Construction Industry:** Ellis Horwood Ltd.

78.   Bramwell, D. (1974), **Computer aided systems in civil engineering using drainage as the prime database:** PhD Thesis, University of Aston.

G.   DATA LIBRARIES AND BILLS OF QUANTITIES

79   Mason, D. (1980), **The limitations of estimators' data bases:** MSc Project Report, Construction Management Course, Loughborough University of Technology.

80.   Gellatly, G.M. (1982), **Development of a structured classification for the Method of Measurement for Road and Bridgeworks for storing estimators' data:** MSc Project Report, Construction Management Course, Loughborough University of Technology.

81.   Fletcher, L. and Moore, T. (1979), **Standard Phraseology for Bills of**

Quantities: George Godwin.

82.   Price, S.G. (1979), **Price's standard method of billing for SMM6:** Quantity Surveyor Weekly.

83.   Department of the Environment (Property Services Agency) (1980), **Enviro BQ System Part 1: Bill item description:** HMSO, London.

84.   Skinner, D.W.H. (1981), **The Contractor's Use of Bills of Quantities.** Occasional Paper No. 24, The Chartered Institute of Building.

H.   COMPUTERS

85.   Gibb, T. (1980), **Humanised computers: the necessity and payoff**: Computers and People, USA.

86.   Shackel, G. (1979), **Man - computer communication: Infotech state of the art report Volume 1: analysis and bibliography:** Infotech.

87.   Stewart, R. (1971), **How computers affect management**: MacMillan.

88.   Stammers, R.B and Patrick, J. (1975), **The Psychology of Training**, in series Essential Psychology, Ed. Herriot, P. Methuen, London.

89.   Hall, D.M. (1981), **An evaluation of user support requirements for a computer aided civil engineering estimating system**: MSc Project Report, Loughborough University of Technology.

96.   Damodran, L., Simpson, A., and Wilson, P. (1981), **Designing Systems for People:** NCC Publications.

91.   Eason, K.D. and Damodran, L. (1979), **Design procedures for user involvement and user support**: HUSAT Memo No. 179, Human Sciences and Advanced Technology Research Centre, Loughborough University of Technology.

92.   Eason, K.D. (1979), **Computer information systems and managerial tasks:** HUSAT Memo No. 175, Human Science and Advanced Technology Centre, Loughborough University of Technology.

93.   Eason, K.D. (1982), **Human factors in Information Technology**: HUSAT Memo No. 252, Human Science and Advanced Technology Centre, Loughborough University of Technology.

94.   Eason, K.D (1982), **The process of introducing Information Technology:** HUSAT Memo No. 248, Human Science and Advanced Technology Centre, Loughborough University of Technology.

Index